Undergraduate Texts in Mathematics

Undergraduate Texts in Mathematics

Abbott: Understanding Analysis.
Anglin: Mathematics: A Concise History and Philosophy.
Readings in Mathematics.
Anglin/Lambek: The Heritage of Thales.
Readings in Mathematics.
Apostol: Introduction to Analytic Number Theory. Second edition.
Armstrong: Basic Topology.
Armstrong: Groups and Symmetry.
Axler: Linear Algebra Done Right. Second edition.
Beardon: Limits: A New Approach to Real Analysis.
Bak/Newman: Complex Analysis. Second edition.
Banchoff/Wermer: Linear Algebra Through Geometry. Second edition.
Berberian: A First Course in Real Analysis.
Bix: Conics and Cubics: A Concrete Introduction to Algebraic Curves.
Brémaud: An Introduction to Probabilistic Modeling.
Bressoud: Factorization and Primality Testing.
Bressoud: Second Year Calculus.
Readings in Mathematics.
Brickman: Mathematical Introduction to Linear Programming and Game Theory.
Browder: Mathematical Analysis: An Introduction.
Buchmann: Introduction to Cryptography.
Buskes/van Rooij: Topological Spaces: From Distance to Neighborhood.
Callahan: The Geometry of Spacetime: An Introduction to Special and General Relavitity.
Carter/van Brunt: The Lebesgue–Stieltjes Integral: A Practical Introduction.
Cederberg: A Course in Modern Geometries. Second edition.
Chambert-Loir: A Field Guide to Algebra.
Childs: A Concrete Introduction to Higher Algebra. Second edition.
Chung/AitSahlia: Elementary Probability Theory: With Stochastic Processes and an Introduction to Mathematical Finance. Fourth edition.
Cox/Little/O'Shea: Ideals, Varieties, and Algorithms. Second edition.
Croom: Basic Concepts of Algebraic Topology.
Curtis: Linear Algebra: An Introductory Approach. Fourth edition.
Daepp/Gorkin: Reading, Writing, and Proving: A Closer Look at Mathematics.
Devlin: The Joy of Sets: Fundamentals of Contemporary Set Theory. Second edition.
Dixmier: General Topology.
Driver: Why Math?
Ebbinghaus/Flum/Thomas: Mathematical Logic. Second edition.
Edgar: Measure, Topology, and Fractal Geometry.
Elaydi: An Introduction to Difference Equations. Second edition.
Erdős/Surányi: Topics in the Theory of Numbers.
Estep: Practical Analysis in One Variable.
Exner: An Accompaniment to Higher Mathematics.
Exner: Inside Calculus.
Fine/Rosenberger: The Fundamental Theory of Algebra.
Fischer: Intermediate Real Analysis.
Flanigan/Kazdan: Calculus Two: Linear and Nonlinear Functions. Second edition.
Fleming: Functions of Several Variables. Second edition.
Foulds: Combinatorial Optimization for Undergraduates.
Foulds: Optimization Techniques: An Introduction.
Franklin: Methods of Mathematical Economics.

(continued after index)

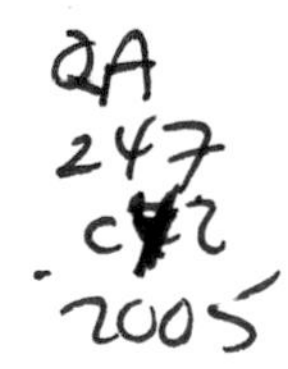

Antoine Chambert-Loir

A Field Guide to Algebra

With 12 Illustrations

Antoine Chambert-Loir
Université de Rennes 1
IRMAR, Campus de Beaulieu
35042 Rennes Cedex
France

Pictured on the cover: Constructing the square root of a real number (Pythagoras's theorem). See page 2 for discussion.

Mathematics Subject Classification (2000): 12-01, 12Fxx, 11Rxx, 13Bxx

Library of Congress Cataloging-in-Publication Data
Chambert-Loir, Antoine.
A field guide to algebra / Antoine Chambert-Loir.
p. cm. — (Undergraduate texts in mathematics, ISSN 0172-6056)
Includes bibliographical references and index.
ISBN 0-387-21428-3 (hardback : alk. paper)
1. Algebraic fields. I. Title. II. Series.
QA247.C48 2004 512′.3—dc22 2004048103

ISBN 0-387-21428-3 Printed on acid-free paper.

Printed in the United States of America. (EBP)

9 8 7 6 5 4 3 2 1 SPIN 10950579

springeronline.com

Then this is a kind of knowledge which legislation may fitly prescribe; and we must endeavour to persuade those who are to be the principal men of our State to go and learn arithmetic, not as amateurs, but they must carry on the study until they see the nature of numbers with the mind only; nor again, like merchants or retail-traders, with a view to buying or selling, but for the sake of their military use, and of the soul herself; and because this will be the easiest way for her to pass from becoming to truth and being.

Plato, *Republic*, Book VII. Engl. transl. by B. Jowett

J'ai fait en analyse plusieurs choses nouvelles.

Évariste Galois, letter to A. Chevalier (29 mai 1832)

Contents

Preface

This is a small book on algebra where the stress is laid on the structure of *fields*, hence its title.

You will hear about equations, both polynomial and differential, and about the algebraic structure of their solutions. For example, it has been known for centuries how to explicitely solve polynomial equations of degree 2 (Babylonians, many centuries ago), 3 (Scipione del Ferro, Tartaglia, Cardan, around 1500 A.D.), and even 4 (Cardan, Ferrari, XVIth century), using only algebraic operations and radicals (nth roots). However, the case of degree 5 remained unsolved until Abel showed in 1826 that a general equation of degree 5 cannot be solved that way.

Soon after that, Galois defined the group of a polynomial equation as the group of permutations of its roots (say, complex roots) that preserve all algebraic identities with rational coefficients satisfied by these roots. Examples of such identities are given by the elementary symmetric polynomials, for it is well known that the coefficients of a polynomial are (up to sign) elementary symmetric polynomials in the roots. In general, all relations are obtained by combining these, but sometimes there are new ones and the group of the equation is smaller than the whole permutation group.

Galois understood how this symmetry group can be used to characterize the solvability of the equation. He defined the notion of *solvable group* and showed that if the group of the equation is solvable, then one can express its roots with radicals, and conversely.

Telling this story will lead us along interesting paths. You will, for example, learn why certain problems of construction by ruler and compass which were posed by the ancient Greeks and remained unsolved for centuries have no solution. On the other hand, you will know why (and maybe discover *how*) one can construct certain regular polygons.

There is an analogous theory for linear differential equations, and we will introduce a similar group of matrices. You will also learn why the explicit computation of certain indefinite integrals, such as $\int \exp(x^2)$, is hopeless.

On the menu are also some theorems from analysis: the transcendance of the number π, the fact that the complex numbers form an algebraically closed field, and also Puiseux's theorem that shows how one can parametrize the roots of polynomial equations, the coefficients of which are allowed to vary.

There are some exercices at the end of each chapter. Please take some time to look at them. There is no better way to feel at ease with the topics in this book. Don't worry, some of them are even easy!

I downloaded the portraits of mathematicians from the *MacTutor History of Mathematics* site, *http://www-groups.dcs.st-andrews.ac.uk/~history/*. I encourage those of you who are interested in History of Mathematics to browse this archive. Reading the books in the bibliography, like the small [4], is also highly recommended. I found the the scans of mathematical stamps at the address *http://jeff560.tripod.com/* — those interested in that subject will be delighted to browse the book [13].

I taught most of this book at École polytechnique (Palaiseau, France). I would like to take the opportunity here to acknowledge all the advice and comments I received from my colleagues, namely, Jean-Michel Bony, Jean Lannes, David Renard and Claude Viterbo. I would also like to thank Sarah Carr for her help in polishing the English translation.

1

Field extensions

We begin with the geometric problem of constructions with ruler and compass. We then introduce the notions of fields, of field extensions, and of algebraic extensions. This will quickly give us the key to the impossibility *of some classical problems. In Chapter 5 we will be able to see how Galois theory gives a definitive criterion allowing us to decide if a geometric construction is, or is not, feasible with ruler and compass.*

1.1 Constructions with ruler and compass

For the Ancient Greeks, the concepts of numbers and of lengths were intimately linked. The problem of *geometric constructions* of remarkable numbers was then naturally posed. Generally, they were allowed to use only ruler and compass, but they sometimes devised ingenious mechanical tools to draw more general curves than lines and circles (*cf.* [4] and the notes of [9]).

Let us give this problem a formal mathematical definition.

Definition 1.1.1. *Let Σ be a set of points in the plane $\mathbf{R}^2$. One says that a point P is* constructible with ruler and compass *from Σ if there is an integer n and a sequence of points $(P_1, \dots, P_n)$ with $P_n = P$ and such that for any $i \in \{1, \dots, n\}$, denoting $\Sigma_i = \Sigma \cup \{P_1, \dots, P_{i-1}\}$, one of the following holds:*

– *there are four points A, B, A' and $B' \in \Sigma_i$ such that P_i is the intersection point of the two nonparallel lines (AB) and $(A'B')$;*

– *there are four points A, B, C, and $D \in \Sigma_i$ such that P_i is one of the (at most) two intersection points of the line (AB) and the circle with center C and radius CD;*

– *there are four points O, M, O' and $M' \in \Sigma_i$ such that P_i is one of the (at most) two intersection points of the distinct circles with, respectively, center O and radius OM, and center O' radius $O'M'$.*

Definition 1.1.2. *Let us consider a subset Σ of $\mathbf{R}$. We say a real number x is* constructible *from Σ if it is the abscissa of a point in the plane which is constructible with ruler and compass from the points $(\xi, 0)$ for $\xi \in \Sigma$. A complex number is defined to be constructible from Σ if its real and imaginary parts are.*

Theorem 1.1.3. *Let Σ be a subset in $\mathbf{R}$ containing 0 and 1. The set $\mathscr{C}_\Sigma$ of all real numbers which are constructible from Σ satisfies the following properties:*

a) *if x and y belong to $\mathscr{C}_\Sigma$, so do $x+y$, $x-y$, xy, and x/y if $y \neq 0$;*
b) *if $x > 0$ belongs to $\mathscr{C}_\Sigma$, then $\sqrt{x} \in \mathscr{C}_\Sigma$.*

Proof. The proof consists of elementary geometrical arguments and can be summed up in a series of figures. Addition and substration are obvious. Stability by multiplication and taking a square root are consequences of Figures 1.1(a) and 1.1(b). Stability by division can also be seen from Figure 1.1(a) for if x and xy are known, the figure shows how to deduce y. □

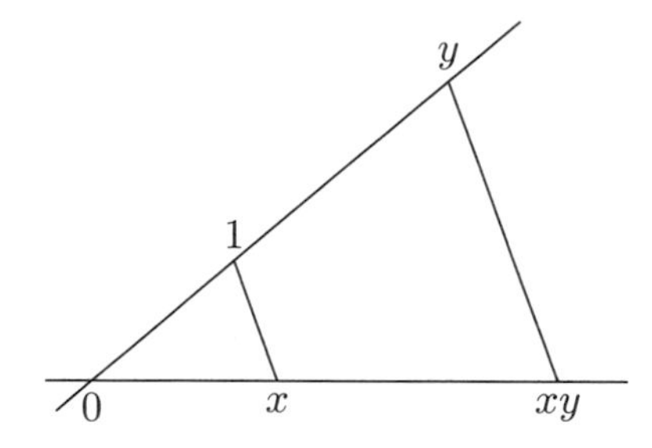

(a) Constructing the product or the ratio of two numbers (Thales's theorem)

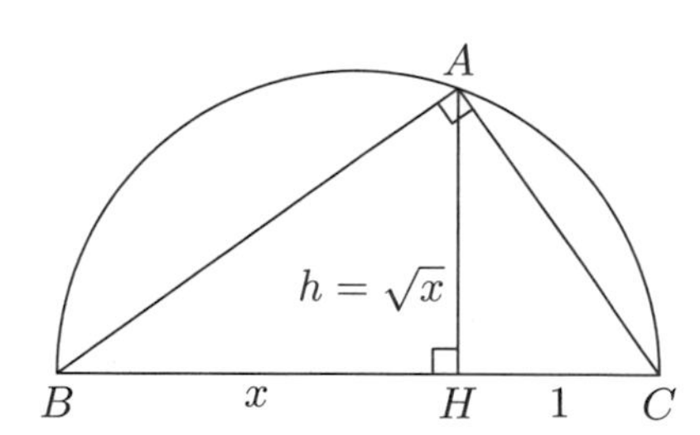

(b) Constructing the square root of a real number (Pythagoras's theorem)

Fig. 1.1. Geometric constructions

Exercise 1.1.4. For these figures to be drawn, it is necessary to be able to construct points outside of the x-axis. You should verify this for yourself. Also construct the line parallel or orthogonal to a given line and going through a fixed point.

In Definition 1.1.1 of a constructible point, the circles are defined by a given center and a given point on the circumference, there are no graduations on the ruler, and the compass closes itself as soon as it does not lie on the plane. Explain, however, how to construct a circle with center a given point and radius the distance between two other points.

Remark 1.1.5. Any construction with ruler and compass can be done with a compass only (theorem of Mohr and Mascheroni). This is a pure geometry result; see, e.g. , [5] for a solution.

1.2 Fields

Definition 1.2.1. *A (commutative)* field *is a set K with two internal laws* $+$ (addition) *and* $\times$ (multiplication) *and two distinct elements* 0 *and* 1 *satisfying the following properties:*

a) $(K, +, 0)$ *is a commutative group*[1];

b) $(K \setminus \{0\}, \times, 1)$ *is a commutative group;*

c) *the law* $\times$ *is distributive with respect to the law* $+$*: for any* a, b *and* c *in* K, $a \times (b + c) = a \times b + a \times c$.

Very often, the product $a \times b$ is denoted ab. One also denotes $K^* = K \setminus \{0\}$.

Examples 1.2.2. *a*) The rational, real and complex numbers form fields, denoted $\mathbf{Q}$, $\mathbf{R}$ and $\mathbf{C}$.

b) The set of numbers (real or complex) which are constructible from $\{0, 1\}$ is a field containing the field of rational numbers.

c) If p is a prime number, the set $\mathbf{Z}/p\mathbf{Z}$ of integers modulo p is a field with p elements.

d) For any field K, the set $K(X)$ of rational functions with coefficients in K, endowed with the usual addition and multiplication, is a field.

e) If Ω is an open domain in $\mathbf{C}$, the set of meromorphic functions on Ω is a field.

Algebraic objects which are defined by the axioms of fields, but without assuming the commutativity of the multiplication, are called *division algebras*. Of course, the law $+$ still has to be commutative.

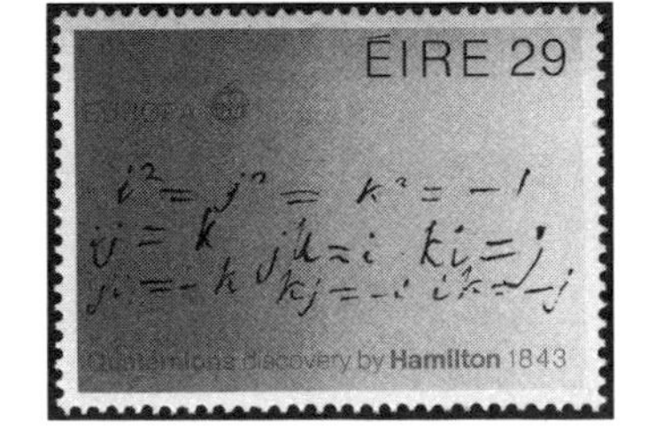

Example 1.2.3. The vector space $\mathbf{H} = \mathbf{R}^4$ whose canonical basis is denoted $\{1, i, j, k\}$

[1] The beginning of Chapter 4 offers a quick reminder of the needed group theory.

admits a unique structure of a division algebra for which the law $+$ is the usual addition of vectors, 1 is the identity for the multiplication, real multiples of 1 commute with any other element, and such the the relations

$$i^2 = j^2 = k^2 = -1, \quad ij = k$$

are satisfied. Other relations follow easily. For example, multiplying the relation $ij = k$ by i on the left, one finds $(-1)j = ik$, hence $ik = -j$.

This is the field of *Hamilton's quaternions*, first discovered by Hamilton.

A *subfield* of a field F is a subset of F containing 0, 1, stable under $+$ and $\times$, so that these laws endow it with the structure of a field.

Definition 1.2.4. *Let K be a field and S be a subset of K. The* field generated by S *in K is the smallest subfield of K containing S.*

This is the set of elements of K of the form

$$\frac{P(s_1, \ldots, s_n)}{Q(s_1, \ldots, s_n)},$$

where $P, Q \in \mathbf{Z}[X_1, \ldots, X_n]$ are polynomials with integer coefficients and $s_1, \ldots, s_n$ are elements in S such that $Q(s_1, \ldots, s_n) \neq 0$. If F is a subfield of K and if $x_1, \ldots, x_n$ belong to K, the subfield of K generated by F and the x_j is also denoted $F(x_1, \ldots, x_n)$.

Exercise 1.2.5. The set of complex numbers of the form $x+iy$ with x, $y \in \mathbf{Q}$ is the subfield of $\mathbf{C}$ generated by i.

A weaker structure than that of a field, but nevertheless very important, is given by a *ring*.

Definition 1.2.6. *A (commutative)* ring *is a set A endowed with two laws $+$ (*addition*) and $\times$ (*multiplication*) and two elements 0 and 1 such that*

- *$(A, +, 0)$ is a commutative group;*
- *the law $\times$ is commutative and associative;*
- *for any $a \in A$, $a \times 1 = 1 \times a = a$;*
- *the law $\times$ is distributive with respect to the law $+$: for any a, b, $c \in A$, $a \times (b + c) = a \times b + a \times c$.*

A subring *of a ring A is a subset of A containing 0 and 1, stable under addition and multiplication, endowing it with a ring structure.*

An element a of a ring A is said to be invertible, *or a* unit, *if there exists an element $b \in A$ with $ab = 1$. If such an element exists, it is necessarily unique and is called the (multiplicative)* inverse *of a.*

Examples 1.2.7. *a*) Fields are rings. More precisely, a field is a ring in which every nonzero element is invertible.

b) The set $\mathbf{Z}$ of integers and the set $\mathbf{Z}/n\mathbf{Z}$ of integers modulo some integer n are rings. The set $\mathbf{Z}$ is a subring of the field of rational numbers.

c) If A is a ring in which $0 = 1$, then $A = \{0\}$ (null ring, of limited interest).

d) If A is a ring, the set $A[X]$ of polynomials with coefficients in A is a ring. The ring A can be identified with the subring of constant polynomials. Details on the algebraic structure of polynomial rings will be given in Section 2.4.

e) If I is an interval in $\mathbf{R}$, the set of continuous functions on I is a ring. The sets of functions which are differentiable, $\mathscr{C}^k$, $\mathscr{C}^\infty$, or analytic form subrings.

f) The set of elements in $\mathbf{C}$ of the form $x + iy$ with x and y in $\mathbf{Z}$ is a subring of $\mathbf{C}$ (*the ring of Gaussian integers*).

g) The set of elements in $\mathbf{H}$ of the form $x1 + yi + zj + tk$ with x, y, z, $t \in \mathbf{Z}$ is a noncommutative subring of the field of quaternions.

Definition 1.2.8. *If A and B are two rings, a* ring homomorphism *is a map $f\colon A \to B$ satisfying the following properties:*

a) *for any a and b in A, $f(a + b) = f(a) + f(b)$;*
b) *for any a and b in A, $f(ab) = f(a)f(b)$;*
c) *$f(0) = 0$, $f(1) = 1$.*

A *field homomorphism* is a ring homomorphism from one field to another. An *isomorphism* (of rings, or of fields) is a bijective homomorphism. The image of a ring morphism $A \to B$ is a subring of B, and the image of a morphism of fields $K \to L$ is a subfield of L.

Definition 1.2.9. *One says a ring A is* integral, *or is an* integral domain, *if $A \neq \{0\}$ and if for any a and $b \in A \setminus \{0\}$, $ab \neq 0$.*

Examples 1.2.10. *a*) Fields and the ring $\mathbf{Z}$ of relative integers are integral domains.

b) A subring of an integral domain is an integral domain.

c) Let n be an integer $\geqslant 2$. The ring $\mathbf{Z}/n\mathbf{Z}$ is an integral domain if and only if n is a prime number.

For any integral domain A, one can define a field K containing (a ring isomorphic to) A so that any element of K is the quotient of two elements of A: the *field of fractions* of A. The same principle is used to construct the rational numbers from the integers. One defines the set K as the set of equivalence classes of the set $\mathscr{F} = A \times (A \setminus \{0\})$ by the equivalence relation

$$(a, b) \sim (c, d) \Leftrightarrow ad = bc.$$

(*Exercise*; show that this is actually an equivalence relation. You will have to use the fact that A is an integral domain.) One denotes by a/b the class of the couple (a, b). Addition and multiplication on K are defined by the usual calculations with fractions, setting

$$\frac{a}{b} + \frac{c}{d} = \frac{ad + bc}{bd} \quad \text{and} \quad \frac{a}{b}\frac{c}{d} = \frac{ac}{bd}.$$

(*Exercise*; check that these operations are well defined, that is $(ad+bc)/bd$ and ab/cd do not depend on the representing fractions a/b and c/d.) Endowed with these two laws, K is a commutative field, its zero is $0/1$ and its 1 is $1/1$. The map $A \to K$ such that a goes to $i(a) = a/1$ is a ring homomorphism. (*Exercise*: check these assertions.) The ring homomorphism i is injective: by definition of the equivalence relation, if $i(a) = a/1 = 0/1$, one deduces $1 \times a = 0 \times 1$, so $a = 0$. There is therefore no harm in identifying an element $a \in A$ with its image $i(a) \in K$. Observe that for any $(a, b) \in \mathscr{F}$,

$$\frac{a}{b} = \frac{a}{1}\frac{1}{b} = i(a)i(b)^{-1}.$$

In other words, any element of K is the quotient of two elements of $i(A)$.

Examples 1.2.11. The field of fractions of the ring $\mathbf{Z}$ is the field of rational numbers. The field of fractions of the ring $K[X]$ of polynomials with coefficients in a field K is the field $K(X)$ of rational functions.

If Ω is a domain in $\mathbf{C}$, the ring of holomorphic functions on Ω is an integral domain (because Ω is connected, this follows from the fact that zeroes of holomorphic functions have no accumulation point in Ω). Its field of fractions is the field of meromorphic functions on Ω. This result is a quite delicate theorem from analysis relying on the ability to construct explicitly a holomorphic function with a prescribed zero-set (Weierstrass products, see, e.g. , [11], Theorem 15.12).

Fraction fields possess an important "*universal property*":

Proposition 1.2.12. *Let A be an integral domain, K its field of fractions. Let E be a field. For any injective ring homomorphism $f\colon A \to E$, there exists a unique field homomorphism $\overline{f}\colon K \to E$ such that $\overline{f}(a) = f(a)$ for $a \in A$.*

Notice that if $a/b = c/d$, then $ad = bc$, so $f(a)f(d) = f(b)f(c)$ and $f(a)/f(b) = f(c)/f(d)$. Therefore, if $x = a/b$ belongs to K, one can set $\overline{f}(x) = f(a)/f(b)$. One then shows that $\overline{f}$ is a field homomorphism. Details of the construction are as tedious as those of the construction of the field of fractions. (*Exercise*...)

One can represent the Proposition visually as a diagram

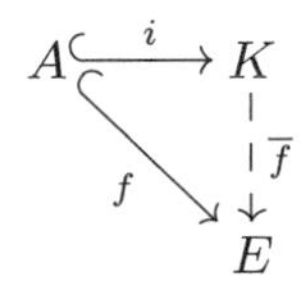

where the dotted arrow $\overline{f}\colon K \to E$ is the one whose existence is precisely asserted by the proposition. A common but pompous terminology for that kind of statement is "universal property."

Lemma 1.2.13. *Let $f\colon A \to B$ be a ring homomorphism. The set $I = f^{-1}(0)$ of elements $a \in A$ with $f(a) = 0$ satisfies the following properties:*

- *$0 \in I$;*
- *if a and $b \in I$, $a + b \in I$;*
- *if $a \in A$ and $b \in I$, $ab \in I$.*

Moreover, f is injective if and only if $I = \{0\}$.

Proof. Left as an exercise! □

Definition 1.2.14. *A subset I of a ring A satisfying the three properties of the preceding lemma is called an* ideal. *If $f\colon A \to B$ is a ring homomorphism, the ideal $f^{-1}(0)$ is called the* kernel *of f and denoted* $\operatorname{Ker} f$.

Proposition/Definition 1.2.15. *Let A be a ring. Then there exists a unique ring homomorphism $f\colon \mathbf{Z} \to A$.*

Assume that f is not injective. If A is an integral domain, the smallest positive element in $\operatorname{Ker} f$ *is a prime number all of whose multiples form* $\operatorname{Ker} f$. *When A is a field, this prime number is called the* characteristic of A.

If f is injective and if A is a field, one says that A is of characteristic zero. *Then f extends to a field homomorphism $g\colon \mathbf{Q} \to A$.*

Proof. Let us first define f. One sets $f(0) = 0$ and $f(1) = 1$. By induction, let $f(n) = f(n-1) + 1$ if $n \geqslant 2$. Finally, if $n \geqslant 1$, define $f(-n) = -f(n)$. As these relations must be verified for any ring homomorphism f, this shows that such a homomorphism $\mathbf{Z} \to A$ is necessarily unique.

Let us show that f is a ring homomorphism, i.e. , let us check the relations $f(m+n) = f(m) + f(n)$ and $f(mn) = f(m)f(n)$. They are in fact true by the same induction argument as the one used to prove that the integers form a ring.

We show that for m and $n \geqslant 0$, one has $f(m+n) = f(m) + f(n)$. This is true if $n = 0$. Moreover, if it is true for an integer n then

$$f(m+(n+1)) = f((m+n)+1) = f(m+n)+1 = f(m)+f(n)+1 = f(m)+f(n+1),$$

so it is again true for $n+1$. By induction, the result is true for any $n \geqslant 0$. If $m \geqslant 0$ and $n < 0$, but $m + n \geqslant 0$, one has

$$f(m+n) - f(m) - f(n) = f(m+n) - f(m) + f(-n)$$
$$= f((m+n) + (-n)) - f(m) = f(m) - f(m) = 0.$$

Other cases are shown similarly.

Let us now show that $f(mn) = f(m)f(n)$ for any m and n. This is obviously true for $n = 0$ and if it is true for an integer n,

$$f(m(n+1)) = f(mn+m) = f(mn) + f(m) = f(m)f(n) + f(m)$$
$$= f(m)(f(n)+1) = f(m)f(n+1),$$

so it is true for $n+1$. By induction, it is true for any $n \geqslant 0$. Finally, if $n \leqslant 0$,

$$f(mn) = f(-m(-n)) = -f(m(-n)) = -f(m)f(-n) = f(m)f(n)$$

as was to be shown.

Assume that A is an integral domain and f is not injective. Consider the smallest positive integer n such that $f(n) = 0$. Since $f(1) = 1 \neq 0$, one has $n \geqslant 2$. If n is not a prime number, one may write $n = ab$ for two positive integers a and b such that $a < n$ and $b < n$. It follows that $0 = f(n) = f(ab) = f(a)f(b)$. As the ring A is an integral domain, one deduces that $f(a) = 0$ or $f(b) = 0$, which contradicts the minimality assumption on n. Any multiple of n goes to 0 by f. Conversely if m is an integer with $f(m) = 0$, Euclidean division of m by n gives us $m = qn + r$ with $0 \leqslant r < n$. Necessarily, $f(r) = f(m - qn) = f(m) - qf(n) = 0$. By minimality $r = 0$ and m is a multiple of n.

If f is injective and if A is a field, the universal property (Prop. 1.2.12) implies that f extends to a homomorphism from $\mathbf{Q}$ to A. □

Remark 1.2.16. Let K be a field of characteristic p and let $f\colon \mathbf{Z} \to K$ be the canonical homomorphism as above. If m and n are two integers belonging to the same congruence class modulo p, $m-n$ is a multiple of p and $f(m-n) = 0$, so $f(m) = f(n)$. The homomorphism $\mathbf{Z} \to K$ induces a natural map $\mathbf{Z}/p\mathbf{Z} \to K$ which is a field homomorphism.

This shows that any field admits one and only one morphism from one of the fields $\mathbf{Z}/p\mathbf{Z}$ (for p prime) or $\mathbf{Q}$, the image of which is called the *prime field*.

Proposition 1.2.17. *Let p be a prime number and let A be a ring such that $p\,1_A = 0_A$ (for example a field of characteristic p). Then any a and b in A satisfy*

$$(a+b)^p = a^p + b^p.$$

It follows that the map $\varphi\colon A \to A$ defined by $\varphi(a) = a^p$ is a ring homomorphism.

Proof. Newton's binomial formula is valid in any (commutative) ring and is written

$$(a+b)^p = a^p + b^p + \sum_{n=1}^{p-1} \binom{p}{n} a^n b^{p-n}.$$

But for any integer n satisfying $1 \leqslant n \leqslant p-1$, the formula $\binom{p}{n} = p!/n!(p-n)!$ shows that $n!(p-n)!\binom{p}{n} = p!$ is a multiple of p. Since p is a prime number, and since $1 \leqslant n \leqslant p-1$, $n!$ and $(p-n)!$ are not multiples of p. It follows that $\binom{p}{n}$ is divisible by p. Therefore, $\binom{p}{n} 1_A = 0$, so $(a+b)^p = a^p + b^p$. □

Definition 1.2.18. *If K is a field of characteristic p, the homomorphism $\varphi\colon K \to K$ given by $x \mapsto x^p$ is called the* Frobenius homomorphism.

1.3 Field extensions

Definition 1.3.1. *Any field homomorphism $j\colon E \to F$ is called a* field extension.

Observe that such a j is always injective, since for any $x \neq 0$,

$$j(x)j(1/x) = j(1) = 1 \neq 0,$$

which implies $j(x) \neq 0$. Most of the time, j is perfectly determined by the context and therefore need not be explicitly stated. One then simply says that F is an extension of E. This is in particular true when $E \subset F$ and j is the inclusion, in which case we will write "let $E \subset F$ be a field extension." If we agree to replace E by its (isomorphic) image under j in F, we can generally think of j as the inclusion map.

If $j\colon E \to F$ is a field extension, F gets naturally enriched with the structure of an E-vector space: the addition law is that from F and the external multiplication $E \times F \to F$ is defined by $e \cdot f = j(e)f$.

Definition 1.3.2. *The* degree *of a field extension $j\colon E \to F$ is the dimension of F as an E-vector space. It is denoted $[F : E]$.*

The extension $j\colon E \to F$ is said to be finite *if $[F : E] \neq +\infty$.*

Remark 1.3.3. There is definitely an abuse of notation in writing $[F : E]$: the homomorphism j does not intervene but the degree depends very much on it! For example, if $E = \mathbf{C}(X)$, $F = \mathbf{C}(Y)$, the extension $j_1\colon E \to F$ defined by $P(X) \mapsto P(Y)$ has degree 1 (it is an isomorphism) but the homomorphism $j_2\colon E \to F$ defined by $P(X) \mapsto P(Y^2)$ has degree 2. When E is a subfield of F, which is the most frequent case, there is no risk of confusion.

Examples 1.3.4. a) The field inclusion $\mathbf{R} \subset \mathbf{C}$ is a finite extension: $\mathbf{C}$ is an $\mathbf{R}$-vector space of dimension 2, the family $\{1, i\}$ being a basis, so $[\mathbf{C} : \mathbf{R}] = 2$.

b) For any field K, the extension $K \subset K(X)$ is not finite. In fact, $K(X)$ contains the infinite free family $\{X^n,\, n \geqslant 0\}$.

Remark 1.3.5. The field inclusion $\mathbf{Q} \subset \mathbf{R}$ is not finite either. Indeed, the product of two countable sets is countable. Since $\mathbf{Q}$ is countable, it follows by induction that any finite-dimensional $\mathbf{Q}$-vector space is countable. But the field of real numbers is not countable, so that $[\mathbf{R} : \mathbf{Q}] = +\infty$. (This argument also shows that the dimension of $\mathbf{R}$ as a $\mathbf{Q}$-vector space is uncountable.)

It is also possible to exhibit infinite families of real numbers which are linearly independent over the rationals. For instance, if α is any transcendental number, then $\{1, \alpha, \alpha^2, \dots\}$ are linearly independent. See also Exercise 1.6 for a more concrete example.

Theorem 1.3.6. *Let $j\colon E \to F$ and $k\colon F \to G$ be two field extensions. Then $(k \circ j)\colon E \to G$ is a finite extension if and only if $j\colon E \to F$ and $k\colon F \to G$ are finite extensions, and one then has the formula*

$$[F : E]\,[G : F] = [G : E].$$

Proof. Choose a basis $x_1, \dots, x_m$ of F as an E-vector space, and a basis $y_1, \dots, y_n$ of G as an F-vector space. Any element $z \in G$ can be written $z = \sum_{i=1}^{n} a_i y_i$ with $a_1, \dots, a_n \in F$. Moreover one may decompose a_i as $a_i = \sum_{j=1}^{m} a_{i,j} x_j$, so that

$$y = \sum_{i=1}^{n} \sum_{j=1}^{m} a_{i,j} x_i y_j.$$

This shows that the family $(x_i y_j)_{\substack{1 \leqslant i \leqslant m \\ 1 \leqslant j \leqslant n}}$ generates G as an E-vector space.

This family is even a basis: let $a_{i,j}$ be elements in E such that $\sum_{i,j} a_{i,j} x_i y_j = 0$. As the family (y_j) is a basis of G as an F-vector space, all elements $\sum_{i=1}^{m} a_{i,j} x_i \in F$ are equal to zero. But now, since the family (x_i) is an E-basis of F, the $a_{i,j}$ have to be zero too.

Consequently,the dimension of G as an E-vector space is equal to $mn = [F : E]\,[G : F]$. □

Definition 1.3.7. *Let $j\colon E \to F$ be a field extension*

An element $x \in F$ is said to be algebraic *over E if there exists a nonzero polynomial $P \in E[X]$ such that $P(x) = 0$. Otherwise, x is called* transcendental.

The extension $E \to F$ is called algebraic *if any element of F is algebraic over E.*

Following historical use, a complex number is said to be algebraic or transcendental if it is such over the field of rational numbers.

Examples 1.3.8. *a*) Consider the field inclusion $\mathbf{R} \subset \mathbf{C}$. An element $z = x + iy$ in $\mathbf{C}$, with x and y in $\mathbf{R}$, satisfies the equation $(z-x)^2 + y^2 = 0$, which shows that z is algebraic over $\mathbf{R}$.

b) The real number $\sqrt{2}$ is algebraic over $\mathbf{Q}$, as is the complex number $\sqrt{2} + i\sqrt[3]{3} + \sqrt[5]{5}$. (*Exercise...*)

c) The real number $\sum_{n=0}^{\infty} 10^{-n!}$ is transcendental (Liouville, 1844); see Exercise 1.2.

d) The set of polynomials with rational coefficients is countable, so that the set of algebraic complex numbers is countable too. Since the set of complex numbers is uncountable, there exist uncountably many transcendental numbers (Cantor, 1874).

e) The real numbers $e \approx 2.718\ldots$, $\pi \approx 3.14159\ldots$ are transcendental (Hermite, 1873, and Lindemann, 1882).

f) It is not known whether or not π is algebraic over the subfield of $\mathbf{R}$ generated by e (whose elements are the $P(e)$ for $P \in \mathbf{Q}(X)$).

Let $j\colon E \to F$ be a field extension and let x be any element in F. The map $\varphi_x\colon E[X] \to F$ associating to any polynomial $P = a_0 + \cdots + a_n X^n$ the element

$$(j(P)(x)) = j(a_0) + j(a_1)x + \cdots + j(a_n)x^n$$

is simultaneously a homomorphism of E-vector spaces and a ring homomorphism. Its image is therefore not only a vector subspace of F but also a subring of F, denoted $E[x]$. It is the subring of F generated by x over E. (If there is no risk of confusion, the element $j(P)(x)$ is also written $P(x)$.) We will soon see (Proposition 1.3.9) that if x is algebraic over E, the subring $E[x]$ in F is actually a field, and so is identified with the subfield $E(x)$ generated by x over E.

Generally, if $x_1, \ldots, x_n$ are elements in F, one denotes by $E[x_1, \ldots, x_n]$ the *subring of F generated by the x_i over E*. It is the set of all $P(x_1, \ldots, x_n) \in F$ for P belonging to $E[X_1, \ldots, X_n]$. The *subfield of F generated by the x_i over E*, denoted $E(x_1, \ldots, x_n)$, is its field of fractions.

The next proposition gives an extremly useful characterization of algebraic elements in terms of the ring $E[x]$.

Proposition 1.3.9. *Let $j\colon E \to F$ be a field extension and let x be an element of F*

a) *If x is transcendental over E, φ_x is injective and $E[x]$ is an infinite-dimensional E-vector space.*

b) *If x is algebraic over E, there exists a unique monic polynomial of minimal degree $P \in E[X]$ such that $P(x) = 0$. Furthermore, P is irreducible and* $\dim_E E[x] = \deg P$. *Moreover, any polynomial $Q \in E[X]$ such that $Q(x) = 0$ is a multiple of P.*

Definition 1.3.10. *This polynomial P is called the* minimal polynomial *of x over E. Its roots (including x) in F are the* conjugates *of x. Its degree is called the* degree *of x over E.*

Recall that a nonconstant polynomial $P \in E[X]$ is said to be *irreducible* if P has no factorization $P = QR$ as the product of two nonconstant polynomials with coefficients in E. A polynomial is said to be *monic* if its leading coefficient is equal to 1. Recall finally that, if $E \to F$ is a field extension, a *root* in F of a polynomial $P \in F[X]$ is an element $x \in F$ such that $P(x) = 0$. By Euclidean division, one can then write $P(X) = (X - x)Q(X)$, for some polynomial $Q \in F[X]$. By induction, the polynomial P has no more roots in F than its degree.

Proof. *a*) Assume that x is transcendental. Then φ_x is injective by definition, so it has to be an isomorphism onto its image, which is $E[x]$. In particular, $\dim_E E[x] = +\infty$.

b) Let $P \in E[X]$ be a monic polynomial with minimal degree such that $P(x) = 0$. Let A be any polynomial in $E[X]$ with $A(x) = 0$. Let $A = PQ + R$ be the Euclidean division of A by P, so that $\deg R < \deg P$. One then has $R(x) = A(x) - P(x)Q(x) = 0$. If R is a nonzero polynomial, with leading coefficient r, the polynomial R/r is monic and its degree is less than that of P, which is a contradiction. It follows that $R = 0$ and A is a multiple of P. (Borrowing from the terminology of Section 2.4, P is the *monic generator of the ideal of polynomials of $E[X]$ vanishing at x.*) Since two monic polynomials dividing each other are necessarily equal, this also implies the uniqueness of such a polynomial P.

Letting $d = \deg P$, the above argument shows that φ_x induces an injective homomorphism $\varphi_{x,d}$ from the E-vector space $E[X]_{<d}$ of polynomials of degree less than d to $E[x]$. By Euclidean division again, $\varphi_{x,d}$ is surjective: let $A \in E[X]$ and write $A = PQ + R$ with $\deg R < d$. Then

$$\varphi_x(A) = A(x) = P(x)Q(x) + R(x) = R(x)$$

belongs to $\operatorname{Im} \varphi_{x,d}$. Therefore $\dim_E E[x] = d$.

It remains to be shown that P is irreducible. But if $P = QR$ for two nonconstant polynomials Q and R in $E[X]$, $Q(x)R(x) = P(x) = 0$, so

$Q(x) = 0$ or $R(x) = 0$. Since Q and R are not constant polynomials, and since $\deg Q + \deg R = \deg P$, $\deg Q < \deg P$ and $\deg R < \deg P$, which again contradicts the minimality assumption for the degree of P. □

Here is the first application.

Corollary 1.3.11. *Every finite extension of fields is algebraic.*

Proof. Let $j \colon E \to F$ be a finite extension of fields. For any $x \in F$, $E[x]$ is an E-vector subspace of F, so has dimension $\leqslant \dim_E F$, and is therefore finite. The preceding proposition implies that x is algebraic over E. □

The following application may be even more surprising.

Theorem 1.3.12. *Let $j \colon E \to F$ be a field extension. Let x and y be two elements of F, which are algebraic over E. Then $x + y$ and xy are algebraic over E. If $x \neq 0$, $1/x$ is algebraic over E and belongs to $E[x]$.*

In particular, any element of $E[x]$ is algebraic over E.

Corollary 1.3.13. *The set of elements of F which are algebraic over E is a subfield of F.*

Proof. Let us introduce the subring $E[x, y]$ of F generated by x and y over E, which is the set of all $P(x, y)$ for $P \in E[X, Y]$. It is a finite-dimensional E-vector space and in fact, if $1, x, \ldots, x^{m-1}$ and $1, y, \ldots, y^{n-1}$ generate $E[x]$ and $E[y]$ respectively, then the set of $x^i y^j$ with $0 \leqslant i < m$ and $0 \leqslant j < n$ spans $E[x, y]$.

This being said, the subrings $E[x + y]$ and $E[xy]$ are both contained in $E[x, y]$. It follows that they are finite dimensional as E-vector spaces and by the preceding proposition, $x + y$ and xy are algebraic over E.

Assuming that $x \neq 0$, let us show that $1/x$ is algebraic over E. As x is algebraic, it satisfies a relation $a_0 + a_1 x + \cdots + a_d x^d = 0$, where the a_i are elements of E, not all zero. Let us divide this relation by x^d. One gets

$$a_0 (1/x)^d + a_1 (1/x)^{d-1} + \cdots + a_d = 0,$$

which proves that $1/x$ is algebraic over E.

I now claim that $1/x$ belongs to $E[x]$. Let r be the smallest integer such that $a_r \neq 0$. One thus has $a_0 = \cdots = a_{r-1} = 0$ and $a_r x^r + \cdots + a_d x^d = 0$, so dividing by $x^r \neq 0$,

$$a_r + a_{r+1} x + \cdots + a_d x^{d-r} = 0.$$

Let us divide this relation once more by $a_r x$. It follows that

$$\frac{1}{x} = -\frac{a_{r+1}}{a_r} - \frac{a_{r+2}}{a_r} x - \cdots - \frac{a_d}{a_r} x^{d-r-1},$$

and $1/x \in E[x]$, as was to be shown. □

Corollary 1.3.14. *An element $x \in F$ is algebraic over E if and only if the ring $E[x]$ is a subfield of F.*

Proof. If the inverse of x, assumed to be nonzero, belongs to $E[x]$, there exists a polynomial $P \in E[X]$ such that $1/x = P(x)$. This implies that x is a root of the nonzero polynomial $1 - XP(X)$, so it is algebraic. Conversely, assume that x is algebraic. For any element $a \in E[x]$, $a \neq 0$, it follows from the preceding theorem that a is algebraic and its inverse in F belongs to $E[a]$. As $E[a] \subset E[x]$, $E[x]$ is a field. (Another proof is given in Exercise 1.1.) □

Remark 1.3.15. Let $j: E \to F$ be a finite extension of fields and let $x \in F$. One saw in the preceding corollaries that x is algebraic over E and $E[x]$ is a subfield of F. We therefore are in the presence of a composed extension $E \to E[x] \to F$ and Theorem 1.3.6 implies that $[F : E] = [F : E[x]]\,[E[x] : E]$. The degree of the extension $E \to E[x]$ is precisely equal to the degree of x. This shows that the degree (over E) of any element of F *divides* the degree $[F : E]$ of the extension $E \to F$.

Another corollary along this line is a "transitivity" property for algebraicity.

Theorem 1.3.16. *Let $j: E \to F$ and $k: F \to G$ be two field extensions. If an element $x \in G$ is algebraic over F and if F is algebraic over E, then x is algebraic over E.*

In particular, if $E \to F$ and $F \to G$ are algebraic extensions, then the composed extension $E \to G$ is also algebraic.

Proof. Let $P \in F[X]$ be the minimal polynomial of x over F, write $P = X^n + a_{n-1}X^{n-1} + \cdots + a_0$. The a_j are in F, so are algebraic over E. By induction, the subring $F_0 = E[a_0, \ldots, a_{n-1}]$ of F is a field and a $E \subset F_0$ is a finite extension. By construction, x is algebraic over F_0 so that the extension $F_0 \to F_0[x]$ is finite. Theorem 1.3.6 implies that the extension $E \to F_0[x]$ is finite, which in turns establishes that x is algebraic over E. □

Remark 1.3.17. If $A \to B$ is a ring homomorphism, one says that B is an *A-algebra*. Besides field extensions (if $E \to F$ is a field extension, F is automatically an E-algebra), particularly important examples of K-algebras are the polynomial rings $K[X_1, \ldots, X_n]$ in n variables $X_1, \ldots, X_n$ over a field K. If A is a ring containing a field K and elements $x_1, \ldots, x_n$ such that $A = K[x_1, \ldots, x_n]$, A is called a *finitely generated K-algebra*.

Proposition 1.3.9 shows in particular that if $A = K[x]$ is a field, then A is algebraic over K. An important theorem proved by Hilbert, often referred to under its German name *Nullstellensatz*, meaning "theorem of the location of zeroes," and which we shall prove in Section 6.8 (Theorem 6.8.1), extends this fact to all finitely generated K-algebras, not only those generated by a single element.

1.4 Some classical impossibilities

I want to show how the results derived in this chapter already allow us to prove that certain geometric constructions are *impossible.*

As the set of constructible numbers is a field, being constructible from $\{0, 1\}$ is equivalent to being constructible from the field $\mathbf{Q}$ of rational numbers.

Theorem 1.4.1 (Wantzel, 1837). *Let E be a subfield of $\mathbf{R}$. A real number x is constructible from E if and only if there exists an integer n and a series of subfields in $\mathbf{R}$,*

$$E = E_0 \subset E_1 \subset \cdots \subset E_n$$

such that for any $i \in \{1, \ldots, n\}$, $[E_i : E_{i-1}] = 2$ and such that $x \in E_n$.

Before proving this, we have to describe the algebraic structure of extensions of degree 2. They are obtained by "adjunction of a square root," which is why they are also called *quadratic extensions.*

Proposition 1.4.2. *Let E be a subfield of $\mathbf{R}$ (more generally any field with characteristic $\neq 2$) and let $j \colon E \to F$ be an extension of degree 2. Then there exists an element $a \in F \setminus E$ such that $a^2 \in E$ and $F = E[a]$.*

Proof. Let x be an element of F which does not belong to E. The family $(1, x)$ is then free over E so is a basis of F as a E-vector space. This implies that the family $(1, x, x^2)$ satisfies a linear relation and there exist three elements a, b, c in E, not all zero, such that $ax^2 + bx + c = 0$. As the family $(1, x)$ is free, $a \neq 0$, which gives us the familiar formula.

$$\left(x + \frac{b}{2a}\right)^2 = \frac{b^2 - 4ac}{4a^2}.$$

Let $\delta = 2ax + b$. Then $\delta^2 = b^2 - 4ac \in E$ is the discriminant of the equation $ax^2 + bx + c$. Since $x = \delta/2a$, $(1, \delta)$ is a basis of F over E. □

Proof of Wantzel's theorem. The proof relies on the form of the equations of lines or circles that intervene in a geometric construction with ruler and compass, and also on the explicit resolution of the equations giving the coordinates of intersection points.

First of all, a line passing through two points $A = (a, b)$ and $A' = (a', b')$ the coordinates of which belong to K has an equation with coefficients in K, namely,

$$\det\begin{pmatrix}1 & 1 & 1\\ x & a & a'\\ y & b & b'\end{pmatrix} = (ab' - a'b) - x(b' - b) + y(a' - a) = 0.$$

Similarly, the circle with radius MM' and center O, where $M = (a, b)$, $M' = (a', b')$ and $O = (a'', b'')$ are points in the plane with coordinates in K, has an equation of the form

$$x^2 + y^2 + Ax + By + C = 0,$$

with A, B, C in K, as is immediately observed by expanding the equation

$$(x - a'')^2 + (y - b'')^2 = (a - a')^2 + (b - b')^2$$

of this circle.

The usual explicit formulae for the coordinates of the intersection point of two intersecting lines are rational expressions in the coefficients of the equations of the lines. Therefore, the intersection point of two nonparallel lines (AA') and (BB') has coordinates in K, if A, A', B, B' are points with coordinates in K.

If P is obtained by intersecting a line and a circle, one writes the polynomial equations of degree $\leqslant 2$,

$$x^2 + y^2 + Ax + By + C = 0 \quad \text{and} \quad Dx + Ey + F = 0,$$

with $A, B, C, D, E, F \in K$. Assuming, for example, $E \neq 0$ and eliminating y, one gets an equation of degree 2 with coefficients in K for x. Let us denote by Δ its discriminant (this is an element of K). Then x belongs to the extension $K(\sqrt{\Delta})$ which has degree $\leqslant 2$ over K, and so does $y = -(Dx + F)/E$.

If P is the intersection point of two circles, one subtracts the equations of the two circles, which brings us back to the case of a circle and a line. (Geometrically speaking, this line is the *radical axis* of the two circles. If the circles meet, it is the line passing through their intersection points.)

By induction on the number of steps, every real number constructible from the subfield E is of the form claimed by the statement of the theorem.

Conversely, if $x \in E_n$, the last field of a chain of quadratic extensions, let us show that x is constructible. By induction, it suffices to show that if $E \subset F$ is such a quadratic extension, every element in F is constructible from E. Proposition 1.4.2 states that there exists an element δ in F such that $F = E[\delta]$ and $\delta^2 \in E$. By Theorem 1.1.3, $\delta = \pm\sqrt{\delta^2}$ is constructible from E, as is any element in $\mathbf{R}$ which is of the form $x + y\delta$. This shows that any element in F is constructible from E. □

Exercise 1.4.3. Extend Wantzel's theorem to complex numbers.

Corollary 1.4.4. *Let E be a subfield in $\mathbf{R}$ and let x be a real number which is constructible from E with ruler and compass. Then x is algebraic over E and its degree is a power of 2.*

Proof. Let $E = E_0 \subset E_1 \subset \cdots \subset E_n \subset \mathbf{R}$ be a chain of quadratic extensions, with $x \subset E_n$. By induction, the multiplicativity of degrees implies that

$$[E_n : E] = [E_n : E_1][E_1 : E_0] = 2[E_n : E_1] = \cdots = 2^n.$$

Considering the composed extension $E \subset E[x] \subset E_n$, one sees that the degree of $E[x]$ over E must divide 2^n, so is necessarily a power of 2. □

We can now prove that certain constructions which were sought in vain for a long time are simply impossible.

Theorem 1.4.5 (Doubling the cube). *The real number $\sqrt[3]{2}$ is not constructible with ruler and compass from $\mathbf{Q}$.*

It is therefore impossible to construct the length of a cube whose volume would be twice that of the unit cube. The legend says that this geometric problem comes from a wish of the Greek god Apollo, who had asked someone to make one of his altars twice as big.

Proof. Set $\alpha = \sqrt[3]{2}$. It suffices to show that the degree of α is not a power of 2. As α is a root of the polynomial $X^3 - 2$, it has degree $\leqslant 3$ and one needs to prove only that $X^3 - 2$ is irreducible over $\mathbf{Q}$, for the degree of α will then be equal to 3.

By Lemma 1.4.9 below, a polynomial of degree 3 is either irreducible or has a root in $\mathbf{Q}$. But the roots of $X^3 - 2$ are α, $\alpha \exp(2i\pi/3)$ and $\alpha \exp(-2i\pi/3)$. Only α is real, so only it might be a rational number. If it were, let us write $\alpha = a/b$ as a fraction in lowest terms. One has $a^3 = 2b^3$ and a is even. Set $a = 2a'$. One then has $b^3 = 4(a')^3$, which shows that b is even too. This contradicts the hypothesis that a and b are coprime. Therefore, α is not a rational number and the polynomial $X^3 - 2$ is irreducible over $\mathbf{Q}$. □

The problem of trisecting an angle is more subtle. From the point with coordinates $(\cos(\alpha), \sin(\alpha))$ on the unit circle, one is asked to construct the point with coordinates $(\cos(\alpha/3), \sin(\alpha/3))$.

Notice that $\sin(\alpha)$ is constructible from the field $\mathbf{Q}(\cos(\alpha))$, because $\sin^2(\alpha) = 1 - \cos^2(\alpha)$. Hence, it is the same for $\cos(\alpha/3)$ to be constructible from $\mathbf{Q}(\cos(\alpha), \sin(\alpha))$ as to be constructible from $\mathbf{Q}(\cos(\alpha))$. Moreover, assuming that $\cos(\alpha/3)$ is constructible from $\mathbf{Q}(\cos(\alpha))$, then $\sin(\alpha/3)$ will be constructible, too. Consequently, *one may trisect the angle α if and only if $\cos(\alpha/3)$ is constructible from $\mathbf{Q}(\cos(\alpha))$.*

As $\cos(3x) = 4\cos^3(x) - 3\cos(x)$, $2\cos(\alpha/3)$ is a root of the polynomial

$$X^3 - 3X - 2\cos(\alpha),$$

the two other roots being $2\cos((\alpha + 2\pi)/3)$ and $2\cos((\alpha + 4\pi)/3)$.

If the polynomial $X^3 - 3X - 2\cos(\alpha)$ is irreducible over $\mathbf{Q}(\cos(\alpha))$, then $\cos(\alpha/3)$ has degree 3 over $\mathbf{Q}(\cos(\alpha))$ and the trisection of angle α cannot be done with ruler and compass. Otherwise, Lemma 1.4.9 implies that has a root in $\mathbf{Q}(\cos(\alpha))$, hence factors as the product of two polynomials in $\mathbf{Q}(\cos(\alpha))[X]$ of degrees 1 and 2. This implies that all of his roots are constructible. We have thus proven the following theorem.

Theorem 1.4.6 (Trisecting an angle). *Let α be a real number. The real number $\cos(\alpha/3)$ is constructible with ruler and compass from $\{0, 1, \cos(\alpha)\}$ if and only if the polynomial $X^3 - 3X - 2\cos(\alpha)$ has a root in the field $\mathbf{Q}(\cos(\alpha))$.*

Example 1.4.7. Let us show that the angle $\pi/9$ cannot be constructed with ruler and compass. As $\cos(\pi/3) = 1/2$, it suffices to prove that the polynomial $P = X^3 - 3X - 1$ has no root in $\mathbf{Q}$. Let us consider such a root, written as a fraction in lowest terms a/b. One has $a^3 - 3ab^2 - b^3 = 0$. If p is a prime number dividing a, it divides $b^3 = a(a^2 - 3b^2)$, and so divides b. As a and b are coprime, $a = \pm 1$. Similarly, if p is a prime number dividing b, it divides $a^3 = b^2(3a + b)$ and it divides a. Therefore, $b = \pm 1$. It follows that only $+1$ and -1 can possibly be rational roots of P. As $P(1) = -3$ and $P(-1) = 1$, P has no root in $\mathbf{Q}$, and so is irreducible over $\mathbf{Q}$.

This shows that one cannot construct with ruler and compass a regular polygon with 9 edges. Later in Chapter 5, Theorem 5.2.2, we will determine which regular polygons are constructible with ruler and compass.

Theorem 1.4.8 (Quadrature of the circle). *The real number $\sqrt{\pi}$ is not constructible.*

In more classical terms, it is impossible to construct with ruler and compass the length of a square whose area would be that of the unit disc.

Proof. If $\sqrt{\pi}$ were constructible, it would be algebraic over $\mathbf{Q}$, and so would π. But F. Lindemann proved in 1882 that π is transcendental; see Theorem 1.6.6. $\square$

The following lemma has been used many times.

Lemma 1.4.9. *Let K be a field. A polynomial $P \in K[X]$ with degree 2 or 3 is irreducible over K if and only if it has no root in K.*

Proof. If P has a root $a \in K$, one can write $P = (X - a)Q$ for some polynomial $Q \in K[X]$ of degree $\leqslant 2$, hence P is not irreductible.

Conversely, assume that P is reducible and write $P = QR$ for two non-constant polynomials Q and R in $K[X]$. As $\deg Q + \deg R = \deg P \in \{2, 3\}$, either $\deg Q$ or $\deg R$ is equal to 1. It necessarily has a root in K, and so does P. □

1.5 Symmetric functions

Recall that the group of permutations (bijections) of the finite set $\{1, \ldots, n\}$ is denoted $\mathfrak{S}_n$. It is a finite group with cardinality $n! = n(n-1)\ldots 2 \cdot 1$.

Definition 1.5.1. *A polynomial $P \in A[X_1, \ldots, X_n]$ in n variables $X_1, \ldots, X_n$ and with coefficients in a ring A is said to be* symmetric *if for any permutation $\sigma \in \mathfrak{S}_n$, one has*

$$P(X_{\sigma(1)}, \ldots, X_{\sigma(n)}) = P(X_1, \ldots, X_n).$$

The most famous examples are the sum $S_1(X) = X_1 + \cdots + X_n$ and the product $S_n(X) = X_1 \ldots X_n$. More generally, one defines the *elementary symmetric polynomials* by

$$S_p(X) = \sum_{1 \leqslant i_1 < \cdots < i_p \leqslant n} X_{i_1} \ldots X_{i_p}, \qquad 1 \leqslant p \leqslant n.$$

It is important to notice that these polynomials are the coefficients of the polynomial $(T - X_1) \ldots (T - X_n)$ (this is a polynomial in the variable T with coefficients in the ring $A[X_1, \ldots, X_n]$). More precisely, one has

$$(T - X_1) \ldots (T - X_n) = T^n - S_1 T^{n-1} + S_2 T^{n-2} + \cdots + (-1)^n S_n.$$

There are many other symmetric polynomials; for example, Newton polynomials

$$N_p(X) = X_1^p + \cdots + X_n^p.$$

These satisfy $N_1 = S_1$,

$$\begin{aligned} N_2(X) &= X_1^2 + \cdots + X_n^2 \\ &= (X_1 + \cdots + X_n)^2 - 2(X_1X_2 + X_1X_3 + \cdots + X_{n-1}X_n) \\ &= S_1^2 - 2S_2, \end{aligned}$$

and in general, $N_p(X)$ can be written as a polynomial with integer coefficients in $S_1(X), \ldots, S_n(X)$.

Proposition 1.5.2. *For any integer $p \geqslant 1$, there is a polynomial with integer coefficients $P_p \in \mathbf{Z}[T_1, \ldots, T_n]$ such that*

$$N_p(X_1, \ldots, X_n) = P_p(S_1(X), \ldots, S_n(X)).$$

Proof. Introduce the polynomial $\Pi = (T - X_1)\dots(T - X_n)$ and let M be its companion matrix. This is the square $n \times n$ matrix

$$\begin{pmatrix} 0 & & & & (-1)^{n-1}S_n \\ 1 & 0 & & & (-1)^{n-2}S_{n-1} \\ & \ddots & \ddots & & \vdots \\ & & & 0 & -S_2 \\ & & & 1 & S_1 \end{pmatrix}$$

with coefficients in the subring $\mathbf{Z}[S_1, \dots, S_n]$ of $\mathbf{Z}[X_1, \dots, X_n]$. The minimal polynomial and the characteristic polynomial of M are both equal to Π. As the characteristic polynomial of M is split in the field $\mathbf{Q}(X_1, \dots, X_n)$, with roots $X_1, \dots, X_n$, the matrix M is conjugate (in that field) to an upper triangular matrix T with diagonal $X_1, \dots, X_n$. By definition, N_p is the trace of T^p, so it is the trace of M^p as well. As M has its coefficients in the ring $\mathbf{Z}[S_1, \dots, S_n]$, so do its powers, and so do their traces. This shows that P_p exists. □

What we have just shown for the Newton polynomials is a general and fundamental fact valid for any symmetric polynomial.

Theorem 1.5.3. *For any symmetric polynomial $P \in A[X_1, \dots, X_n]$, there exists a unique polynomial $Q \in A[Y_1, \dots, Y_n]$ such that*

$$P(X_1, \dots, X_n) = Q(S_1(X), \dots, S_n(X)).$$

Proof. The existence of Q is shown by induction first on the number of variables n, and then on the degree of P. If $n = 1$, one has $S_1 = X_1$ and one sets $Q = P$. If $\deg P = 0$, P is constant and one chooses for Q this constant. Assume that the result is satisfied in $(n-1)$ variables, and that it holds in degrees $< m$ for n variables, and let P be any symmetric polynomial of degree m in $X_1, \dots, X_n$. The polynomial P_0 in $(n-1)$ variables $X_1, \dots, X_{n-1}$ defined by

$$P_0(X_1, \dots, X_{n-1}) = P(X_1, \dots, X_{n-1}, 0)$$

is symmetric. By induction, there is a polynomial

$$Q_0 \in A[Y_1, \dots, Y_{n-1}]$$

such that

$$P_0(X_1, \dots, X_{n-1}) = Q_0(S_1(X), \dots, S_{n-1}(X)).$$

In this last formula, there appear symmetric polynomials in $(n-1)$ variables, but it is readily verified that, letting the number of variables equal the exponent,

$$S_p^{(n-1)}(X_1,\ldots,X_{n-1}) = S_p^{(n)}(X_1,\ldots,X_{n-1},0)$$

and more generally,

$$S_p^{(n)}(X_1,\ldots,X_n) = S_p^{(n-1)}(X_1,\ldots,X_{n-1}) + X_n S_{p-1}^{(n-1)}(X_1,\ldots,X_{n-1}).$$

Then

$$P_1(X) = P(X_1,\ldots,X_n) - Q_0(S_1(X),\ldots,S_{n-1}(X))$$

is a symmetric polynomial which becomes the null polynomial when X_n is replaced by 0. The coefficient of any monomial $X_1^{i_1}\ldots X_n^{i_n}$ is zero as soon as $i_n = 0$. Since P_1 is symmetric, the coefficient of $X_1^{i_1}\ldots X_n^{i_n}$ is also null as soon as any one of the i_j is zero. Therefore, any nonzero monomial in P_1 is a multiple of $S_n = X_1\ldots X_n$ and so is P_1. Let us write $P_1 = S_n P_2$ for some $P_2 \in A[X_1,\ldots,X_n]$. The polynomial P_2 is still symmetric but has degree $< m$. By induction, one may write $P_2 = Q_2(S_1,\ldots,S_n)$. Finally,

$$P(X) = Q_0(S_1,\ldots,S_n) + P_1(X) = Q_0(S_1,\ldots,S_n) + S_n Q_2(S_1,\ldots,S_n)$$

and it suffices to set $Q = Q_0 + Y_n Q_2$.

Let us now show uniqueness. It suffices to show that for any polynomial $Q \in A[Y_1,\ldots,Y_n]$ satisfying $Q(S_1,\ldots,S_n) = 0$, one has $Q = 0$. This is obvious for $n = 1$. Assume that the uniqueness property is established for $(n-1)$ variables and let us show the result for n variables by induction on the degree of Q. Setting X_n equal to 0, one has in particular

$$\begin{aligned} 0 &= Q(S_1(X_1,\ldots,X_{n-1},0),\ldots,S_n(X_1,\ldots,X_{n-1},0)) \\ &= Q(S_1^{(n-1)},\ldots,S_{n-1}^{(n-1)},0), \end{aligned}$$

which implies by induction that $Q(Y_1,\ldots,Y_{n-1},0) = 0$. The polynomial Q is therefore a multiple of Y_n and one concludes by induction on the degree of Q. □

An important symmetric polynomial is the *discriminant*:

$$D = \prod_{i<j}(X_i - X_j)^2.$$

To prove that it is indeed symmetric, it may be more convenient to write it as

$$D = (-1)^{n(n-1)/2}\prod_{i\neq j}(X_i - X_j)$$

and to observe that for any permutation $\sigma \in \mathfrak{S}_n$, the map $(i,j) \mapsto (\sigma(i),\sigma(j))$ is a bijection from the set of ordered pairs of distinct integers in $\{1,\ldots,n\}$ onto itself. This shows that for any $\sigma \in \mathfrak{S}_n$,

$$\begin{aligned} D(X_{\sigma(1)},\dots,X_{\sigma(n)}) &= (-1)^{n(n-1)/2}\prod_{i\neq j}(X_{\sigma(i)}-X_{\sigma(j)}) \\ &= (-1)^{n(n-1)/2}\prod_{i\neq j}(X_i-X_j) \\ &= D(X_1,\dots,X_n), \end{aligned}$$

hence D is symmetric.

1.6 Appendix: Transcendence of e and π

We show in this section that e and π are transcendental. As these real numbers are not defined by algebra but really by analysis, it should not be surprising that the proof involves analytic tools, united in the following lemma.

Lemma 1.6.1. *Let f be a polynomial with real coefficients; let m be its degree. For any complex number z, the integral*

$$I(f;z) = \int_0^1 ze^{z(1-u)} f(zu)\,du$$

satisfies

$$I(f;z) = e^z \sum_{j=0}^m f^{(j)}(0) - \sum_{j=0}^m f^{(j)}(z).$$

Moreover, the following upper bound is valid:

$$|I(f;z)| \leqslant |z|\,e^{|z|} \sup_{u\in[0;1]} |f(zu)|\,.$$

Proof. By integration by parts on the definition of $I(f;z)$, one finds

$$\begin{aligned} I(f;z) &= \Big[-e^{z(1-u)}f(zu)\Big]_0^1 + \int_0^1 e^{z(1-u)}zf'(zu)\,du \\ &= -f(z) + e^z f(0) + I(f';z), \end{aligned}$$

hence the result by induction on the degree of f. The upper bound for $|I(f;z)|$ follows from the fact that, for any $u \in [0,1]$, one has

$$\left|ze^{z(1-u)}f(zu)\right| \leqslant |z|\,e^{|z|} \sup_{u\in[0,1]} |f(zu)|\,,$$

by integration over $[0,1]$. □

Lemma 1.6.2. *Let f be a polynomial with integer coefficients. For any integer $n \geqslant 0$, there exists a polynomial f_n with integer coefficients such that $f^{(n)} = n!f_n$.*

Proof. By linearity, it suffices to prove this for $f = X^m$. In this case, $f^{(n)} = m(m-1)\dots(m-n+1)X^{m-n}$. Therefore, the polynomial $f_n = \binom{m}{n}X^{m-n}$ has integer coefficients and $f^{(n)} = n!f_n$. □

Theorem 1.6.3 (Hermite). *e is a transcendental number.*

Proof. Let us reason ad absurdum. If e were algebraic, there would exist integers $a_0, \dots, a_n$, not all equal to 0, such that

$$a_0 + a_1 e + \dots + a_n e^n = 0.$$

Dividing this relation by a power of e, if necessary, we may assume that $a_0 \neq 0$. Let p be a prime number (fixed for the moment, but later we will take the limit as p goes to infinity).

Let $f(X) = X^{p-1}(X-1)^p \dots (X-n)^p$ and introduce

Charles Hermite (1822–1901)

$$J_p = a_0 I(f;0) + a_1 I(f;1) + \dots + a_n I(f;n).$$

It follows that

$$J_p = -\sum_{k=0}^{n} a_k \sum_{j=0}^{np+p-1} f^{(j)}(k),$$

which in particular is an integer.

By Lemma 1.6.1, there exists a real number c such that $|J_p| \leqslant c^p$ for all p.

Moreover, if $k \in \{1, \dots, n\}$, then k is a root of f with multiplicity p, so that $f^{(j)}(k) = 0$ for $j < p$. If $j \geqslant p$, $f^{(j)}(k)$ is also divisible by $p!$, by Lemma 1.6.2 above. Observe, however, that $k = 0$ is a root of f with multiplicity exactly $p-1$. Therefore $f^{(j)}(0) = 0$ for $j < p-1$, and $f^{(j)}(0)$ is still a multiple of $p!$ for $j \geqslant p$, but

$$f^{(p-1)}(0) = (p-1)!(-1)^p \dots (-n)^p = (-1)^{np}(p-1)!(n!)^p.$$

Consequently, there exists an integer N_p such that

$$J_p = (-1)^{np+1} a_0 (p-1)!(n!)^p + p!N_p.$$

In particular, if $p > n$ and p does not divide a_0, then $J_p/(p-1)!$ is an integer which is nonzero modulo p, and therefore $J_p/(p-1)! \neq 0$. But any nonzero integer has absolute value at least equal to 1, implying $|J_p| \geqslant (p-1)!$. Since $a_0 \neq 0$, large prime numbers p do not divide a_0 and, for those, one gets the inequality $c^p \geqslant |J_p| \geqslant (p-1)!$. However, this contradicts Stirling's formula

$$p! \sim p^p e^{-p} \sqrt{2\pi p}$$

when p goes to infinity. □

Let us now prove that π is transcendental. If f is any nonzero polynomial and $g\colon \mathbf{C} \to \mathbf{C}$ any function, we will denote by

$$\sum_{f(\alpha)=0} g(\alpha)$$

the sum $g(\alpha_1) + \cdots + g(\alpha_n)$, the α_j being the roots of f repeated according to their multiplicities.

Lemma 1.6.4. *Let f be a polynomial with integer coefficients and with leading coefficient c. Then, for all $n \geqslant 0$,*

$$c^n \sum_{f(\alpha)=0} \alpha^n \in \mathbf{Z}.$$

Proof. Let m be the degree of f and let A denote the companion matrix with minimal polynomial f/c. The eigenvalues of A are the roots of f, with the same multiplicities. It follows that the eigenvalues of cA are the $c\alpha$, for $f(\alpha) = 0$, and those of $c^n A^n$ are the $c^n \alpha^n$, with $f(\alpha) = 0$. Hence, $c^n \sum_{f(\alpha)=0} \alpha^n$ is precisely equal to the trace of $c^n A^n$. By hypothesis, cA is a matrix with integer coefficients. It follows that $c^n A^n$ has also integer coefficients and its trace is an integer. $\square$

Proposition 1.6.5. *Let f be any polynomial with integer coefficients such that $f(0) \neq 0$ and assume the sum $\sum_{f(\alpha)=0} e^\alpha$ is an integer. Then this sum is equal to zero.*

Proof. Let $S = \sum_{f(\alpha)=0} e^\alpha$ and let us assume that S is a nonzero integer. Let c be the leading coefficient of f.

Let p be a prime number, and define $g(X) = X^{p-1} f^p(X)$. This is a polynomial with integer coefficients and with degree $m = p(1 + \deg f) - 1$. Finally, let

$$J_p = \sum_{f(\alpha)=0} I(g; \alpha).$$

The upper bound for I given by Lemma 1.6.1 implies that there exists a real number $M > 0$ such that, for any p,

$$|J_p| \leqslant M^p.$$

It also follows from Lemma 1.6.1 that

$$J_p = S\left(\sum_n g^{(n)}(0)\right) - \sum_n \left(\sum_{f(\alpha)=0} g^{(n)}(\alpha)\right).$$

If $f(\alpha) = 0$, α is root of g with multiplicity p, so that $g^{(n)}(\alpha) = 0$ for $n < p$. Moreover, if $n \geqslant p$, Lemma 1.6.2 asserts that $g_n = g^{(n)}/p!$ is a polynomial with integer coefficients and degree $m - n \leqslant m - p$. The preceding lemma now implies that there exists an integer A_p such that

$$c^{m-p} \sum_n \sum_{f(\alpha)=0} g^{(n)}(\alpha) = p! A_p.$$

Furthermore, $g^{(n)}(0) = 0$ for $n < p - 1$, is divisible by $p!$ for $n \geqslant p$, but

$$g^{(p-1)}(0) = (p-1)! f(0)^p.$$

This shows that there exists an integer B_p such that

$$\sum_n g^{(n)}(0) = (p-1)! f(0)^p + p! B_p.$$

Finally,

$$J_p = (p-1)! f(0)^p S + p!(c^{p-m} A_p + S B_p).$$

In particular,

$$\frac{c^{m-p}}{(p-1)!} J_p = c^{m-p} f(0)^p S + p\big(A_p + c^{m-p} S B_p\big)$$

is an integer. Moreover, if the prime number p does not divide $cf(0)S$, then this integer is not divisible by p. It is therefore nonzero, and has absolute value at least 1. This gives us the lower bound

$$|J_p| \geqslant (p-1)! c^{p-m} = (p-1)! c^{1-p \deg f}.$$

Since $cSf(0) \neq 0$, any large enough prime number satisfies the last inequality. The above upper bound for J_p then implies that

$$(p-1)! c^{1-p \deg f} \leqslant M^p,$$

which again contradicts Stirling's formula when p goes to infinity. □

Theorem 1.6.6 (Lindemann). *π is transcendental.*

Proof. If π were algebraic, $i\pi$ would be algebraic too. Let then f be an irreducible polynomial with integer coefficients such that $f(i\pi) = 0$. Let us denote its roots by $\alpha_1, \ldots, \alpha_n$. From Euler's Equation $1 + e^{i\pi} = 0$, one deduces that

$$\prod_{f(\alpha)=0} (1 + e^{\alpha}) = (1 + e^{\alpha_1}) \ldots (1 + e^{\alpha_n}) = 0.$$

Let us expand this equality. One gets

$$\sum_{\varepsilon\in\{0,1\}^n} \exp(\sum \varepsilon_j \alpha_j) = 0.$$

But now the sums $\sum \varepsilon_j \alpha_j = 0$ are the roots of the polynomial

$$F_0(X) = \prod_{\varepsilon\in\{0,1\}^n} (X - \sum_j \varepsilon_j \alpha_j),$$

whose coefficients obviously can be expressed with symmetric polynomials in the α_j. By Theorem 1.5.3 on symmetric polynomials, the coefficients of F_0 are polynomials with integer coefficients in the elementary symmetric polynomials of the α_j, so in the coefficients of f. It follows that F_0 has rational coefficients. Consequently, there exists an integer $N \geqslant 0$ such that $NF_0 \in \mathbf{Z}[X]$. One has $F_0(0) = 0$ (corresponding to $\varepsilon_j = 0$ for all j); let $q \geqslant 1$ be the multiplicity of this root 0 and let us set $F = NF_0/X^q$. Then F is a polynomial with integer coefficients and $F(0) \neq 0$. Moreover, one has

$$0 = \sum_{\varepsilon\in\{0,1\}^n} \exp(\sum \varepsilon_j \alpha_j) = q + \sum_{F(\beta)=0} e^{\beta}.$$

Since $q \neq 0$, this contradicts Prop. 1.6.5. □

Exercises

Exercise 1.1. **a)** Let A be integral domain. If A is finite, show that A is a field. Examples?

b) Let F be an integral domain and let $E \subset F$ be a subring of E which is a field. Assume moreover that F is a finite-dimensional E-vector space. Show that F is a field.

Exercise 1.2 (Liouville's criterion). Let α be a complex number. Assume that α is algebraic over $\mathbf{Q}$ and let d be its degree.

a) Show that there exists a polynomial $P \in \mathbf{Z}[X]$ with degree d such that $P(\alpha) = 0$ and $P'(\alpha) \neq 0$.

b) Using a), show that there exists a real number $c > 0$ such that for any $(p, q) \in \mathbf{Z} \times \mathbf{N}^*$, one has

$$\left| \alpha - \frac{p}{q} \right| \geqslant \frac{c}{q^d}.$$

c) Show that the real number

$$\alpha = \sum_{n=1}^{\infty} 10^{-n!}$$

is transcendental over $\mathbf{Q}$. (Liouville, 1844). A real number, the transcendental nature of which can be established this way, is said to be a *Liouville number.* The set of all Liouville numbers is uncountable but has measure 0 in $\mathbf{R}$. One also knows that π and e are not Liouville numbers.

Exercise 1.3. Let $\mathbf{C}(z)$ denote the field of rational functions with complex coefficients. Let Ω be any domain in $\mathbf{C}$ and let $\mathscr{M}(\Omega)$ denote the field of meromorphic functions on Ω. Let $j\colon \mathbf{C}(z) \to \mathscr{M}(\Omega)$ be the natural field homomorphisms associating to a rational function the corresponding meromorphic function on Ω.

a) Let $f \in \mathbf{C}(z)$ be any nonconstant rational function without poles in Ω. Show that $\exp(f) \in \mathscr{M}(\Omega)$ does not belong to $\mathbf{C}(z)$.

b) It f is any element in $\mathbf{C}(z) \setminus \mathbf{C}$, show that $\exp(f)$ is transcendental over $\mathbf{C}(z)$. (By contradiction, if N denotes the degree of $\exp(f)$ over $\mathbf{C}(z)$, differentiate with respect to z a nontrivial algebraic relation $\sum\limits_{n=0}^{N} p_n(z)\exp(nf(z)) = 0$. Conclude that $\exp(Nf(z)) \in \mathbf{C}(z)$.)

c) If $f_1, \dots, f_n$ are distinct nonconstant elements of $\mathbf{C}(z)$, show that $\exp(f_1), \dots,$ $\exp(f_n)$ are linearly independent over $\mathbf{C}(z)$. (Prove this by induction on n. Consider a relation $\sum\limits_{i=1}^{n} p_i(z)\exp(f_i(z)) = 0$. If $p_n \neq 0$, divide it by $p_n(z)$ and differentiate.)

Exercise 1.4. **a)** Let $P = X^n + a_{n-1}X^{n-1} + \cdots + a_0$ be a polynomial of degree n with complex coefficients. Show that any root $z \in \mathbf{C}$ of P satisfies

$$|z| \leqslant 1 + |a_0| + \cdots + |a_{n-1}|.$$

b) Let $f\colon \mathbf{C} \to \mathbf{C}$ be an *entire* function, that is, a holomorphic function defined on the whole complex plane. Assume that f is algebraic over the field $\mathbf{C}(z)$ of rational functions. Show that there is an integer $n \geqslant 0$ and a real number c such that, for any $z \in \mathbf{C}$,

$$|f(z)| \leqslant c(1 + |z|^n).$$

c) (*continued*) Let $f(z) = \sum\limits_{j=0}^{\infty} c_j z^j$ be the Taylor expansion of f. Prove that the function g defined by $g(z) = \sum\limits_{j=0}^{\infty} c_{j+n} z^j$ is entire and bounded. Deduce from Liouville's theorem on bounded entire functions that f is a polynomial.

Exercise 1.5. Let P be a monic polynomial in $\mathbf{Z}[X]$. If $a \in \mathbf{Q}$ is a root of P, show that $a \in \mathbf{Z}$.

Exercise 1.6. **a)** Let $E \subset F$ be a quadratic extension. Consider $x \in F \setminus E$ such that $x^2 \in E$. Let $a \in E$. If a is a square in F, prove that either a is a square in E, or ax^2 is a square in E.

b) Let $p_1, \dots, p_n$ be distinct prime numbers. One considers the following two properties:

(a_n) the field $\mathbf{Q}(\sqrt{p_1}, \dots, \sqrt{p_n})$ has degree 2^n over $\mathbf{Q}$;

(b_n) an element $x \in \mathbf{Q}$ is a square in $\mathbf{Q}(\sqrt{p_1}, \dots, \sqrt{p_n})$ if and only if there exists a subset $I \subset \{1, \dots, n\}$ such that $x \prod_{i \in I} p_i$ is a square in $\mathbf{Q}$.

Show that (a_n) and (b_n) together imply (a_{n+1}), and that (a_n) and (b_{n-1}) imply (b_n). Deduce by induction on n that both hold for any integer n.

c) Show that the square roots $\sqrt{2}, \sqrt{3}, \sqrt{5}, \sqrt{7}, \dots$ of all prime numbers are linearly independent over the rational numbers.

Exercise 1.7. Let p be any prime number and consider the polynomial $P = X^n + X + p$, with $n \geqslant 2$.

a) Assume that $p \neq 2$. If z denotes a complex root of P, show that $|z| > 1$.

b) Still for $p \neq 2$, prove that P is irreducible in $\mathbf{Z}[X]$.

c) Assume that $p = 2$. If n is even, show that P is irreducible in $\mathbf{Z}[X]$. If n is odd, show that $X + 1$ divides P and that $P(X)/(X + 1)$ is irreducible in $\mathbf{Z}[X]$.

d) More generally, any polynomial $P = a_0 + a_1 X + \dots + a_n X^n \in \mathbf{Z}[X]$, such that $|a_0|$ is a prime number greater than $|a_1| + \dots + |a_n|$, is irreducible.

Exercise 1.8. Let $P = X^n + a_{n-1}X^{n-1} + \dots + a_0$ be any monic polynomial in $\mathbf{Z}[X]$ such that $a_0 \neq 0$ and

$$|a_{n-1}| > 1 + |a_{n-2}| + \dots + |a_0| .$$

a) Using Rouché's theorem in the theory of complex variables, show that P has exactly one complex root of absolute value $\geqslant 1$.

b) Show that P is irreducible in $\mathbf{Z}[X]$ (*Perron's theorem*).

Exercise 1.9 (Content of a polynomial). If P is any polynomial with integer coefficients, define the *content* of P as the greatest common divisor of its coefficients, denoted $\operatorname{ct}(P)$.

a) Let P and Q be two polynomials in $\mathbf{Z}[X]$. Let p be a prime number which divides all coefficients of PQ. Show by reduction modulo p that p divides either $\operatorname{ct}(P)$ or $\operatorname{ct}(Q)$.

b) Show for any polynomials P and Q in $\mathbf{Z}[X]$ that $\operatorname{ct}(PQ) = \operatorname{ct}(P)\operatorname{ct}(Q)$.

c) Let P be a monic polynomial in $\mathbf{Z}[X]$ and let Q be a monic polynomial in $\mathbf{Q}[X]$ which divides P in $\mathbf{Q}[X]$. Show that Q has integer coefficients.

Exercise 1.10 (Eisenstein's criterion for irreducibility). *Let p be a prime number and let $A = a_n X^n + \dots + a_0$ be any polynomial with integer coefficients such that a) p divides $a_0, \dots, a_{n-1}$; b) p does not divide a_n; c) p^2 does not divide a_0. Then A is irreducible in $\mathbf{Q}[X]$.*

The proof is ad absurdum assuming that A is reducible in $\mathbf{Q}[X]$.

a) Using part a) of the previous exercise, show that there exist nonconstant polynomials B and $C \in \mathbf{Z}[X]$ such that $A = BC$.

b) Let us denote $B = b_d X^d + \dots + b_0$. Reducing modulo p, show that p divides b_0, $\dots, b_{d-1}$.

c) Show that p^2 divides a_0. This is a contradiction.

d) Show that the polynomial

$$\frac{X^p - 1}{X - 1} = X^{p-1} + \cdots + 1$$

is irreducible in $\mathbf{Q}[X]$. (Apply a change of variables $X = Y + 1$.)

Exercise 1.11. Show that the set of constructible complex numbers is a subfield of $\mathbf{C}$ which is stable under taking square roots.

Exercise 1.12. In a 1833 book devoted to geometry, *Traité du compas (Traité élémentaire de tous les traits servant aux Arts et Métiers et à la construction des Bâtiments)* by Zacharie [14], one can find the following construction.

> *Construct a regular heptagon, that is a figure with seven equal sides.* From any point draw a circle; draw the diameter AB, divide this diameter in seven equal parts, at points 1, 2, 3, 4, 5, 6, 7; construct two circles with centers A and B, each with radius AB, meeting at the point C; from the intersection point C, draw the line $C5$ which you will extend until it meets the circle at a point D; draw the line BD, it will be the side of the heptagon; with the compass, mark the length of line BD on the circle, at points E, F, G, H, I and you will get the heptagon.

Draw a picture and explain where the construction goes wrong.

Exercise 1.13. Let P denote the polynomial $X^4 - X - 1$.

a) Show that it has exactly two real roots. One denotes them x_1 and x_2. Let x_3 and x_4 be the two other complex roots.

b) Show that P is irreducible over $\mathbf{Q}$. (You may either reduce modulo 2 or observe that P has exactly one root of absolute value less than one.)

c) Let $P(X) = (X^2 + aX + b)(X^2 + cX + d)$, where a, b, c and d are to be determined. Express b, c and d in terms of a. Then construct a polynomial Q of degree 3 such that, a being fixed, this system has a solution if and only if $Q(a^2) = 0$.

d) Show that Q is irreducible over $\mathbf{Q}$.

e) Prove that x_1 and x_2 cannot both be constructible with ruler and compass. (In fact, it will follow from Exercise 5.4 that neither of them can be.)

Exercise 1.14 (Newton's formulae). **a)** Prove the following formulae relating Newton sums and elementary symmetric polynomials in $\mathbf{Z}[X_1, \ldots, X_n]$:

$$\text{if } m \leqslant n, \qquad N_m - N_{m-1}S_1 + \cdots + (-1)^{m-1}N_1S_{m-1} + (-1)^m mS_m = 0$$

$$\text{if } m > n, \qquad N_m - N_{m-1}S_1 + \cdots + (-1)^n N_{m-n}S_n = 0.$$

b) Deduce from this that any symmetric polynomial in $\mathbf{Q}[X_1, \ldots, X_n]$ can be written uniquely as a polynomial with rational coefficients in the Newton sums $N_1, \ldots, N_n$.

c) Is this still true in a field of characteristic $p > 0$?

Exercise 1.15. Let $(G,+)$ be a finite abelian group. One says that an element $g \in G$ has *order* d if d is the smallest integer $\geqslant 1$ such that $dg = 0$.

a) Let g and h be two elements of G, with orders m and n. If m and n are coprime, show that $g + h$ has order mn.

b) More generally, if G possesses two elements with orders m and n, show that there is an element in G with order $\operatorname{l.c.m.}(m,n)$.

c) Show that there exists an integer $d \geqslant 1$ and an element $g \in G$ such that a) g has order d; b) for any $h \in G$, $dh = 0$.

Exercise 1.16. Let E be a commutative field and let G be a finite subgroup of E^*. Show that G is a cyclic group. (Consider a couple (d,g) as in Exercise 1.15, c), and show that g generates G.)

Exercise 1.17. Let $j \colon K \to E$ be a field extension, $x_1, \dots, x_n$ elements in E. Show that the following properties are equivalent:

a) the x_i are algebraic over K;
b) $K[x_1, \dots, x_n]$ is finite dimensional over K;
c) $K[x_1, \dots, x_n]$ is a field;
d) $K(x_1, \dots, x_n)$ is finite dimensional over K.

(That (c) implies (d) requires Hilbert's Nullstellensatz, Theorem 6.8.1.)

Exercise 1.18. Define the *degree*, the *weight* and the *partial degree* of a monomial $X_1^{i_1} \dots X_n^{i_n}$ as the quantities $i_1 + \dots + i_n$, $i_1 + 2i_2 + \dots + ni_n$ and $\max(i_1, \dots, i_n)$. The degree, the weight and the partial degree of a polynomial P, denoted respectively $\deg(P)$, $w(P)$ and $\delta(P)$, are by definition the maximum of the degrees, weights and partial degrees of its nonzero monomials.

a) Compute the degrees, weights and partial degrees of the elementary symmetric polynomials $S_1, \dots, S_n$.

b) Let $P \in \mathbf{Z}[X_1, \dots, X_n]$ be a symmetric polynomial. By Theorem 1.5.3, there exists a unique polynomial $Q \in \mathbf{Z}[X_1, \dots, X_n]$ such that $P = Q(S_1, \dots, S_n)$.

By returning to the inductive proof of Theorem 1.5.3, show that $\deg(P) = w(Q)$ and $\delta(P) = \deg(Q)$.

2

Roots

In the first chapter, the emphasis was on given numbers, and we were led to look at the equations of which they are solutions. In this chapter, we switch roles and look at polynomial equations and their eventual roots. Generalizing the construction of the field of complex numbers from the real numbers, we show how to create roots of a polynomial which does not have enough of them in a given field.

2.1 Ring of remainders

Let K be a field and let P be a nonconstant polynomial with coefficients in K. We denote its degree by d. Endow the vector space $E \subset K[X]$ of polynomials with degree $< d$ with a ring structure in the following way:

– addition, with its identity element 0, is given by the vector space structure;

– the identity element for the multiplication is the constant polynomial 1;

– if A and B are two polynomials in E, one defines the multiplication $A * B$ as the remainder in the Euclidean division of the polynomial AB by the polynomial P.

Let us show that this actually defines a ring. First of all, it is obvious that $(E, +, 0)$ is an abelian group. The internal law $*$ is obviously commutative; moreover, $1 * A = A * 1$ is the remainder in the Euclidean division of A by P, so is equal to A since $\deg A < d = \deg P$. This shows that 1 is the identity for the law $*$. To show associativity, consider the equations $AB = PQ_1 + A * B$ and $(A * B)C = PQ_2 + (A * B) * C$ obtained from Euclidean division, so that

$$\begin{aligned} ABC &= (PQ_1 + A * B)C = PQ_1C + PQ_2 + (A * B) * C \\ &= P(Q_1C + Q_2) + (A * B) * C. \end{aligned}$$

This shows that $(A*B)*C$ is the remainder in the Euclidean division of ABC by P. Similarly, $A*(B*C)$ is the remainder in the Euclidean division of ABC by P, so is equal to $(A*B)*C$. The law $*$ is thus associative. The distributivity is shown in the same way: considering the equations $AB = PQ_1 + A*B$ and $AC = PQ_2 + A*C$ obtained by Euclidean division, one deduces that

$$A(B+C) = AB + AC = P(Q_1 + Q_2) + A*B + A*C.$$

Hence, the remainder in the Euclidean division of $A(B+C)$ by P is equal to $A*B + A*C$, which gives us the equality $A*(B+C) = A*B + A*C$.

Let us also remark that mapping the element $a \in K$ to the constant polynomial $a \in E$ defines a ring homomorphism $K \to E$.

Definition 2.1.1. *The ring that we just constructed is denoted $K[X]/(P)$.*

This is *the ring of remainders of Euclidean divisions* by P. As soon as we are familiar with this new ring, we will drop the symbol $*$ and just denote multiplication as usual.

Proposition 2.1.2. *Let P be a nonconstant polynomial in $K[X]$. The following properties are equivalent:*

a) *the ring $K[X]/(P)$ is a field;*
b) *the ring $K[X]/(P)$ is an integral domain;*
c) *the polynomial P is irreducible in $K[X]$.*

Proof. Implication *a)*⇒*b)* is obvious. Assume *b)*. If $P = QR$ in $K[X]$, for two polynomials Q and R with degrees $< \deg P$, one has $Q*R = 0$ in $K[X]/(P)$, which contradicts the hypothesis that $K[X]/(P)$ is an integral domain, so P is irreducible in $K[X]$, hence *c)*. Finally, assume *c)*. Let A be a nonzero element of $K[X]/(P)$, viewed as a polynomial of degree $< \deg P$; we have to show that A has an inverse in $K[X]/(P)$. Since P is irreducible and A is not a multiple of P, they are coprime and Bézout's relation (Corollary 2.4.2) gives us two polynomials U and V in $K[X]$ such that $UA + VP = 1$. If $U = PQ + U_1$ is the Euclidean division of U by P, then $U_1 * A = 1$, which shows that A is invertible in the ring $K[X]/(P)$. □

Let P be an irreducible polynomial in $K[X]$ and consider the field extension $j\colon K \to K[X]/(P)$ that we have just defined. Let x denote the polynomial X viewed as an element in $K[X]/(P)$. By construction, for every polynomial $A \in K[X]$, $A(x)$ is the remainder in the Euclidean division of the polynomial $A(X)$ by the polynomial P. In particular, $P(x) = 0$. In other words, *we just defined an extension of the field K in which the polynomial P has a root.* The next theorem claims that this is actually the "best" way of doing it. In fact, this ring $K[x]/(P)$ satisfies a *universal property*.

Theorem 2.1.3. *Let K be a field and let P be a polynomial in $K[X]$. Let us denote the ring $K[X]/(P)$ by A and let $j\colon K \to A$ be the canonical ring homomorphism. Now let $i\colon K \to B$ be a ring homomorphism and y an element in B such that $P(y) = 0$. Then there exists a* unique *ring homomorphism $f\colon A \to B$ such that $f \circ j = i$ and $f(x) = y$.*

This is sometimes represented by a diagram

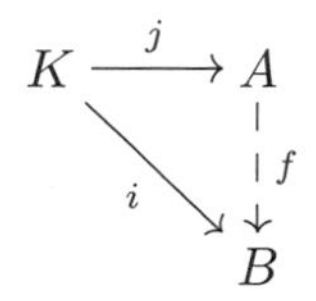

where the dotted arrow $f\colon A \to B$ is the one whose existence is claimed by the theorem.

The idea of the proof is not complicated, yet requires understanding of what we did during the construction of $K[X]/(P)$. We started with the ring $K[X]$ in which we have a new element $x = X$, but this satisfies no relation at all and it is not a root of P. Then we changed the rules in a clever way by imposing $P(x) = 0$. Some of the consequences of $P(x)$ vanishing come from Euclidean division: if $A = QP + B$, then the relation $P(x) = 0$ forces $A(x) = B(x)$. A posteriori, the validity of the given construction actually means that all consequences are obtained from Euclidean divisions.

Proof. If $f(x) = y$, one must have $f(Q(x)) = Q(y)$ for any polynomial $Q \in K[X]$, and in particular for any polynomial with degree $< \deg P$. That shows that there exists at most one homomorphism f satisfying $f \circ j = i$, and that if it actually exists, it has to be given by the map

$$f\colon K[X]/(P) \to B, \quad Q(X) \mapsto Q(y).$$

(Recall that an element of $K[X]/(P)$ *is* really a polynomial with degree $<$ $\deg P$.) Let us define f in this way. We now have to prove that f is a ring homomorphism. It is obviously a morphism of vector spaces and it satisfies $f \circ j = i$. Moreover, if Q and R are two polynomials in $K[X]$ with degrees $< \deg P$, let us write the Euclidean division of QR by P, say, $QR = PS + Q * R$. Then, since $P(y) = 0$ in B,

$$\begin{aligned} f(Q * R) &= (Q * R)(y) = (Q * R)(y) + P(y)S(y) \\ &= (QR)(y) = Q(y)R(y) = f(Q)f(R), \end{aligned}$$

which shows that f is a ring homomorphism. □

To sum up the construction of this section, let us introduce a definition.

Definition 2.1.4. *If* $i\colon E \to F$ *and* $j\colon E \to F'$ *are two extensions of a field* E*, a* homomorphism of extensions *from* F' *to* F *is a field homomorphism* $f\colon F' \to F$ *such that* $f \circ j = i$*.*

A bijective homomorphism of extensions is called an *isomorphism.*

Theorem 2.1.5. *Let* K *be a field and let* P *be a irreducible polynomial with coefficients in* K*. There exists a finite field extension* $K \to K_1$ *and a root* x *of* P *in* K_1 *such that*

a) $K_1 = K[x]$*;*

b) *If* $K \to L$ *is a field extension, the set of morphisms of extensions from* K_1 *to* L *is in bijection with the set of roots of* P *in* L*, this bijection being given by* $f \mapsto f(x)$*.*

2.2 Splitting extensions

Definition 2.2.1. *Let* K *be a field and let* P *be a nonconstant polynomial in* $K[X]$*. A* splitting extension *of* P *is a field extension* $j\colon K \to E$ *such that:*

a) *over* E*,* P *can be decomposed as a product of linear factors; explicitely, if* d *is the degree of* P *and if* c *denotes its leading coefficient, then there exist* $x_1, \dots, x_d$ *in* E *such that* $P = c\prod_{i=1}^{d}(X - x_i)$*;*

b) *as a field,* E *is generated by the* x_i*, that is,* $E = K(x_1, \dots, x_d)$*.*

In other words, a splitting extension of an irreducible polynomial P is an extension which contains all of "the" roots of P (this is condition *a*) and which is "minimal" for that property (this is condition *b*).

Theorem 2.2.2. *Let* K *be a field and let* P *be a nonconstant polynomial in* $K[X]$*.*

a) *There is a splitting extension for* P*.*

b) *Any two such extensions are isomorphic: if* $j\colon K \to E$ *and* $j'\colon K \to E'$ *are two splitting extensions of* P*, there exists an isomorphism of fields* $f\colon E \to E'$ *such that* $f \circ j = j'$*.*

Proof. Let us begin with a very simple remark: let $K \to E$ be a splitting extension of P and let α be a root of P in E. This allows us to write $P = (X - \alpha)Q$, where Q is a polynomial with coefficients in $K[\alpha]$. It is then clear that E is a splitting extension of the polynomial Q over the field $K[\alpha]$. This remark gives the idea of the proof of the theorem: if we know how to construct $K[\alpha]$, we will obtain a splitting extension E by induction. And we know precisely how to define a field $K[\alpha]$; that was the main result of the last section.

Let us now show *a)* and *b)* by induction on the degree of P. If $\deg P = 1$, it suffices to let $E = K$. Let $Q \in K[X]$ be an irreducible factor of P. By Theorem 2.1.5, there exists an extension $K \to K_1$ and an element $x_1 \in K_1$ such that *a)* $Q(x_1) = 0$; *b)* $K_1 = K[x_1]$. Then let P_1 be the quotient of P by $X - x_1$ in the ring $K_1[X]$. By induction, the polynomial P_1 admits a splitting extension over K_1, say $K_1 \to E$. The composed extension $K \to E$ is a field extension in which P is a product of factors of degree 1. Moreover, if $x_2, \dots, x_d$ are the roots of P_1 in E (set $d = \deg P$),

$$E = K_1(x_2, \dots, x_d) = K(x_1)(x_2, \dots, x_d) = K(x_1, \dots, x_d),$$

so that E is generated by the x_i over K. Hence E is a splitting extension of P over K.

Let $K \to E'$ be another splitting extension of P and let us define an isomorphism of extensions from E to E'. By hypothesis, the chosen irreducible factor Q of P has a root x_1' in E'. By Theorem 2.1.5, there exists a homomorphism of extensions $f_1 \colon K_1 \to K_1'$, where $K_1' = K[x_1']$ is the subfield of E' generated by x_1'. As f_1 is surjective, it has to be an isomorphism and this isomorphism maps the polynomial P_1 to the polynomial $P_1' = P/(X - x_1')$. The composite extension $K_1 \xrightarrow{\sim} K_1' \to E'$ is therefore a splitting extension of the polynomial $P/(X - x_1)$. By induction, the two extensions $K_1 \to E$ and $K_1 \to E'$ are isomorphic and there exists an isomorphism $f \colon E \to E'$ extending the isomorphism $f_1 \colon K_1 \to K_1'$. □

2.3 Algebraically closed fields; algebraic closure

Definition 2.3.1. *One says that a field K is* algebraically closed *if any nonconstant polynomial of $K[X]$ has a root in K.*

By induction on the degree, we see that this statement is equivalent to saying that *any polynomial is split in K*. The constructions of this chapter also show that a field is algebraically closed if and only if *it has no nontrivial algebraic extensions* (that is, if $j \colon K \to E$ is an algebraic extension, j is a isomorphism). One direction is clear. If K is algebraically closed and if $j \colon K \to E$ is an algebraic extension, let x be an element of E, P its minimal polynomial. By hypothesis, P is split in K: there exist elements $x_1, \dots, x_n$ in K such that $P = (X - x_1) \dots (X - x_n)$. Since $P(x) = 0$, x is one of the x_i (more precisely one of the $j(x_i)$). This shows that j is surjective and hence an isomorphism. For the other direction, let P be a nonconstant polynomial in $K[X]$ and let Q be an irreducible factor of P. We showed that the ring $K[X]/(Q)$ is an algebraic extension of K with degree $\deg Q$. Since K has no nontrivial algebraic extensions, $\deg Q = 1$, so that Q has a root in K, and so does P.

Definition 2.3.2. *An* algebraic closure *of a field K is an algebraic extension $j \colon K \to \Omega$, where Ω is an algebraically closed field.*

Theorem 2.3.3 (Steinitz, 1910). *Every field has an algebraic closure; two algebraic closures of a field are isomorphic.*

There are two types of algebraic closures: those that one can see, like the algebraic closure of the field of real numbers (which is the field of complex numbers), and those which are constructed by a transfinite procedure, as in the general proof of the existence of an algebraic closure.

Theorem 2.3.4. *The field $\mathbf{C}$ of complex numbers is algebraically closed.*

Despite its famous name, "the fundamental theorem of algebra," this is really a theorem from analysis. Let me offer you three proofs. The first one is short and frankly analytic. The second one looks as if it were algebraic, but analysis is hidden in the use of the "intermediate value theorem." The third one comes from topology.

First proof. Let $P \in \mathbf{C}[X]$ be a nonconstant polynomial with no root in $\mathbf{C}$. Let us write it $P = a_n X^n + \cdots + a_0$, with $a_n \neq 0$ and $n \geqslant 1$. Then, for any $z \in \mathbf{C}$ such that $|z| > 1$, one has

$$
\begin{aligned}
|P(z)| &\geqslant |a_n|\,|z|^n - (|a_0| + \cdots + |a_{n-1}|)\,|z|^{n-1} \\
&\geqslant |z^n| \left(|a_n| - \frac{1}{|z|}(|a_0| + \cdots + |a_{n-1}|) \right).
\end{aligned}
$$

In particular, $|P(z)|$ goes to $+\infty$ when $|z| \to \infty$. It follows that the function $1/P$ is bounded on $\mathbf{C}$ and holomorphic everywhere (P does not vanish). By Liouville's theorem, it is constant, hence we have a contradiction. □

Second proof. Let $P \in \mathbf{C}[X]$ be a nonconstant polynomial. Observe that the polynomial $Q(X) = P(X)\overline{P}(X)$ has real coefficients. If we show that it has a complex root z, then either $P(z) = 0$, or $P(\overline{z}) = \overline{\overline{P}(z)} = 0$, so that P also has a complex root. Therefore it suffices to show that every nonconstant polynomial $P \in \mathbf{R}[X]$ has a complex root, which we will prove by induction on the greatest power of 2, $\nu_2(P)$, which divides the degree of P.

If this power is 0, that is, if $\deg P$ is odd, the limits of $P(x)$ when $x \to \pm\infty$ are $+\infty$ and $-\infty$ (depending on the sign of the leading coefficient of P). It follows from the intermediate value theorem that P has a real root.

Assume the result is established for polynomials P such that $\nu_2(P) < n$ and let P be a polynomial in $\mathbf{R}[X]$ with $\nu_2(P) = n$. Let Ω be an extension of $\mathbf{C}$ in which P is split and let us denote its roots by $(\xi_i)_{1\leqslant i\leqslant \deg P}$. Let c be a real number. For $1 \leqslant i < j \leqslant \deg P$, set $z_{i,j;c} = \xi_i + \xi_j + c\xi_i\xi_j$ and let us introduce the monic polynomial $Q \in \Omega[X]$ whose roots are the $z_{i,j;c}$. First of all, one has $\deg Q = \deg P(\deg P - 1)/2$, hence $\nu_2(Q) = \nu_2(P) - 1$. Moreover, Q has *real coefficients.* Indeed, these coefficients are given by polynomials with integer coefficients in the ξ_i, and these polynomials are invariant under every permutation of the variables. It follows from Theorem 1.5.3 on symmetric polynomials that the coefficients of Q can be expressed as polynomials with real coefficients in the elementary symmetric polynomials of $\xi_1, \dots, \xi_{\deg P}$, that is, in the coefficients of P. In particular, the coefficients of Q are real numbers. By induction, Q has at least one root in $\mathbf{C}$.

This is true for every value of c. As $\mathbf{R}$ is infinite, there exists at least one pair (i, j) and two real numbers $c \neq c'$ such that $\xi_i+\xi_j+c\xi_i\xi_j$ and $\xi_i+\xi_j+c'\xi_i\xi_j$ both are complex numbers, from which we deduce that $a = \xi_i+\xi_j$ and $b = \xi_i\xi_j$ belong to $\mathbf{C}$. They are roots of the polynomial $R = X^2 - aX + b$, whose discriminant $\Delta = a^2 - 4b$ is a complex number. If we show that Δ is a square in $\mathbf{C}$, it will follow that the two roots of R, namely ξ_i and ξ_j, are complex numbers.

Let $\Delta = p + iq$. The equation $(x + iy)^2 = \Delta$ is equivalent to the equations

$$x^2 - y^2 = p \quad \text{and} \quad 2xy = q,$$

hence $(x^2 + y^2)^2 = p^2 + q^2$ and $x^2 + y^2 = \sqrt{p^2 + q^2}$. One obtains for x^2 and y^2 the following (nonnegative) values:

$$x^2 = \frac{1}{2}\big(p + \sqrt{p^2 + q^2}\big) \quad \text{and} \quad y^2 = \frac{1}{2}\big(- p + \sqrt{p^2 + q^2}\big),$$

hence values for x and y, by accounting their signs so that $q = xy/2$.

This shows that ξ_i and ξ_j are complex numbers and consequently that the initial polynomial P has a root in $\mathbf{C}$. By induction, the theorem is proved. □

Third proof. Again let P be any nonconstant polynomial with coefficients in $\mathbf{C}$. If $z \in \mathbf{C}$, we will denote by $\nu(z)$ the cardinality of the finite set $P^{-1}(z)$. The goal is to show that $\nu(0) > 0$ and we will in fact show that $\nu(z) > 0$ for every $z \in \mathbf{C}$.

Let $\Delta \subset \mathbf{C}$ be the set of $z \in \mathbf{C}$ such that $P'(z) = 0$, $U = \mathbf{C} \setminus \Delta$ and $V = \mathbf{C} \setminus P(\Delta)$. The sets U and V are the complementary subsets of finite sets in $\mathbf{C}$, so they are open and connected (*exercise*).

If $u = x + iy$ and $P(u) = A(x, y) + iB(x, y)$, one deduces easily (for example, from Cauchy's formulae in the theory of analytic functions) that

$$|P'(u)|^2 = \det \begin{pmatrix} \partial A/\partial x & \partial B/\partial x \\ \partial A/\partial y & \partial B/\partial y \end{pmatrix}.$$

Therefore, the implicit function theorem for functions $\mathbf{R}^2 \to \mathbf{R}^2$ implies that for any $u \in \mathbf{C}$ with $P'(u) \neq 0$, P defines a diffeomorphism from a neighborhood of u to a neighborhood of $P(u)$.

Now let $z \in V$. For every $u \in P^{-1}(z)$, one has $P'(u) \neq 0$, so that there exists a neighborhoods W_u and Ω_u of z such that P induces a diffeomorphism $W_u \to \Omega_u$. Let $\Omega = \bigcap_{u \in P^{-1}(z)} \Omega_u$; this is a neighborhood of z and any $w \in \Omega$ has at least $\nu(z)$ preimages by P, one in each W_u, $u \in P^{-1}(z)$. In particular, the set V^+ of $z \in V$ such that $\nu(z) > 0$ is *open* in V.

But it is also closed: let (z_j) be any sequence of points in V with $\nu(z_j) > 0$ such that $z_j \to z \in V$. Let us choose for every j an element $u_j \in \mathbf{C}$ such that $z_j = P(u_j)$. Since the sequence (z_j) is bounded and $|P(u)| \to +\infty$ when $|u| \to +\infty$, the sequence (u_j) is bounded too. It thus has a limit point $u \in \mathbf{C}$. Since P defines a continuous function, $P(u)$ is also a limit point of the sequence $(P(u_j))$. Necessarily $P(u) = z$, and hence $\nu(z) > 0$. This shows that V^+ is closed in V.

As V is connected, the nonempty subset V^+ cannot be both open and closed unless it is equal to all of V. In other words, $\nu(z) > 0$ for every $z \in V$.

If $z \notin V$, there exists by definition $u \in \Delta$ such that $P(u) = z$ and $\nu(z) > 0$. Finally, $\nu(z) > 0$ for every $z \in \mathbf{C}$. □

From an algebraically closed field, it is easy to construct an algebraic closure for any of its subfields.

Proposition 2.3.5. *Let Ω be an algebraically closed field and let K be a subfield in Ω. Let $\overline{K}$ be the set of elements in Ω which are algebraic over K. Then $K \subset \overline{K}$ is an algebraic closure of K.*

For instance, the set of algebraic numbers in $\mathbf{C}$ is an algebraic closure of $\mathbf{Q}$.

Proof. The extension $K \subset \overline{K}$ is algebraic by construction, for every element in $\overline{K}$ is algebraic over K.

Let $P \in \overline{K}[X]$ be a nonconstant polynomial and let us show that it has a root in $\overline{K}$. As $\overline{K} \subset \Omega$ and as Ω is algebraically closed, P has a root x in Ω. The element x is algebraic over $\overline{K}$ and since $\overline{K}$ is algebraic over K, x is also algebraic over K (Theorem 1.3.16). Therefore $x \in \overline{K}$ and P has a root in $\overline{K}$, as was to be shown. □

The proof of Steinitz's theorem is not very illuminating and relies upon a "transfinite induction" argument, hence requires the axiom of choice as soon as the field is not countable! We have shown how to add the roots of one polynomial, and all we have to do is to add roots for all of them, which requires the set of polynomials to be well-ordered.

Proof of Steinitz's theorem. Let K be a field whose algebraic closure is to be constructed. We are going to define an algebraic extension $K \to \Omega$ of K in which every polynomial of $K[X]$ is split. It will follow that Ω is an algebraic closure of K. Let P be a polynomial $P = X^n + a_{n-1}X^{n-1} + \cdots + a_0$ with coefficients in Ω. We have to show that P has a root in Ω. We may assume that P is irreducible. Since every coefficient a_i is algebraic over K, the subfield $L = K[a_0, \dots, a_{n-1}] \subset \Omega$ which they generate is a finite algebraic extension of K. Necessarily P is irreducible in $L[X]$. Let us then introduce the finite algebraic extension $L \to L[X]/(P)$, in which P has a root α, with minimal polynomial P. Since L is algebraic over K, α is algebraic over K. Let Q denote its minimal polynomial in $K[X]$. As $Q(\alpha) = 0$, Q is a multiple of P in $L[X]$. By construction, Q is split in Ω. It follows that P is split too, so it has a root in Ω.

The method of constructing Ω consists of patiently "adding" the roots of every irreducible polynomial in $K[X]$. To that aim, endow the set $\mathscr{E}$ of all irreducible polynomials with a *well-ordering* $\prec$, that is, a total ordering such that every nonempty subset of $\mathscr{E}$ admits a least element. The standard ordering on $\mathbf{N}$ is a well-ordering and the existence of a well-ordering on any set is equivalent to the axiom of choice, or to Zorn's lemma. If K is countable, the set of all irreducible polynomials with coefficients in K is also countable and enumerating them gives us a well-ordering.

Once a set is well-ordered, the induction principle can be stated and proved in quite the same way as the classical induction over the integers. Let $(X, \prec)$ be a well-ordered set and let $\mathscr{P}$ be a property of elements in X. Assume that the following assertion holds (induction hypothesis):

> "Let $x \in X$; if, for every $y \in X$, $y \prec x$, $\mathscr{P}(y)$ is true, then $\mathscr{P}(x)$ is true."

Then $\mathscr{P}(x)$ is true for every $x \in X$. (Otherwise, the set of $x \in X$ such that $\mathscr{P}(x)$ does not hold admits a smallest element x_0. By definition, for every $y \prec x_0$, $\mathscr{P}(y)$ is true. By the assertion within quotes, $\mathscr{P}(x_0)$ is true; thus we have a contradiction. The first step of the induction, i.e. , checking $\mathscr{P}$ for the minimal element of X, follows by applying the induction hypothesis with $x = \min(X)$.)

Let us now show the existence of a family of algebraic extensions $j_P \colon K \to \Omega_P$, for $P \in \mathscr{E}$, in which P is split, and of homomorphisms $j_P^Q \colon \Omega_Q \to \Omega_P$ where P and Q are two polynomials in $\mathscr{E}$ with $Q \prec P$, satisfying $j_P = j_P^Q \circ j_Q$. (This means that Ω_P is an extension not only of K but also of all Ω_Q for $Q \prec P$.)

To show this by induction, two constructions are needed, where $P \in \mathscr{E}$.

– The first, which I do not want to do formally, is an *inductive limit* $\Omega_{\prec P}$ of all extensions Ω_Q with $Q \prec P$. This is essentially the union of all these fields; to compute in $\Omega_{\prec P}$, one chooses some Ω_Q where everything is defined and one computes there. Using the homomorphisms $j_{Q'}^Q$, one sees that the output of those calculations is essentially independent of the field where they were done. One has moreover homomorphisms $j_{\prec P}^Q \colon \Omega_Q \to \Omega_{\prec P}$.

– The second consists in adding to the field $\Omega_{\prec P}$ all roots of P; one defines Ω_P as a splitting extension of the polynomial P over the field $\Omega_{\prec P}$, hence a field homomorphism $j_P^{\prec P}\colon \Omega_{\prec P} \to \Omega_P$ which, composed with $j_{\prec P}^Q$, gives us the homomorphisms $j_P^Q\colon \Omega_Q \to \Omega_P$ we sought.

Once these (Ω_P, j_P^Q) are shown to exist, we define Ω as the inductive limit of all Ω_P.

To prove that two algebraic closures are isomorphic, we will use a theorem from the next chapter. Let $K \to \Omega'$ be an algebraic closure of K. We want to show that there exists a K-homomorphism from the algebraic closure we just constructed Ω to Ω'. Let us show by induction that there exists, for every $P \in \mathscr{E}$, a K-homomorphism $\alpha_P\colon \Omega_P \to \Omega'$ such that $\alpha_P \circ j_P^Q = \alpha_Q$ if $Q \prec P$. Let us now fix P. The homomorphisms $\alpha_Q\colon \Omega_Q \to \Omega'$ for $Q \prec P$, $Q \neq P$, define a field homomorphism from $\Omega_{\prec P}$, which is the inductive limit of the Ω_Q for $Q \prec P$, to Ω'. Applying Theorem 3.1.6 to the field Ω_P (which is a splitting extension of the polynomial P over $\Omega_{\prec P}$), there exists a morphism of field extensions $\Omega_P \to \Omega'$ which extends the morphism $\Omega_{\prec P} \to \Omega'$.

Together, the α_P define a K-homomorphism $\alpha\colon \Omega \to \Omega'$. Like any field homomorphism, α is injective. Let us show it is surjective. Let x be any element in Ω'. By definition, x is algebraic over K so let $P \in K[X]$ be its minimal polynomial. As Ω is an algebraic closure of K, P is split in Ω. Writing $P = \prod_{i=1}^{n}(X - x_i)$ in $\Omega[X]$, one then has

$$0 = P(x) = \prod_{i=1}^{n}(x - \alpha(x_i))$$

so that x is one of the $\alpha(x_i)$ and α is surjective, q.e.d. □

2.4 Appendix: Structure of polynomial rings

Recall that an *ideal* of a ring A is a subgroup $I \subset A$ such that for any $a \in A$ and $b \in I$, $ab \in I$. If a is any element of A, the *principal ideal* generated by a is the set of all ab for $b \in A$. We denote it aA or (a). Conversely, we say that a is a *generator* of the ideal (a).

Theorem 2.4.1. *For any ideal I in $K[X]$, there exists a polynomial $P \in K[X]$ such that $I = (P)$.*

An integral domain in which every ideal is a principal ideal is called a *principal ideal ring*.

Proof. We essentially have to redo the argument of Proposition 1.3.9 of which this theorem is a particular case: just take for I the set of all polynomials $P \in K[X]$ such that $P(x) = 0$. If $I = \{0\}$, simply set $P = 0$. Otherwise, let $d \geqslant 0$ be the smallest degree of a nonzero element in I and let $P \in I$ be a polynomial of degree d. Since I is an ideal, $PQ \in I$ for any $Q \in K[X]$, so $(P) \subset I$. Conversely, let A be an element in I and consider the Euclidean division $A = PQ + R$ of A by P. One has $PQ \in I$, so that $R = A - PQ$

belongs to I. By definition, $\deg R < \deg P = d$. The definition of d implies $R = 0$ and $A = PQ \in (P)$. □

Also notice that a nonzero ideal in $K[X]$ has many generators. However, if P and Q are two generators of a nonzero ideal, then there exists a constant $\lambda \in K^*$ such that $P = \lambda Q$. Indeed, P and Q divide each other; writing $P = RQ$ and $Q = SP$ implies that R and S are nonzero constants. Consequently, every nonzero ideal of $K[X]$ has a unique generator which is a monic polynomial.

Corollary 2.4.2 (Bézout's theorem for polynomials). *Let A and B be two polynomials. The set $I = (A, B)$ consisting of all $AP + BQ$ with P, $Q \in K[X]$ is an ideal in $K[X]$. If D is a generator of this ideal, then*

a) *there exist U and $V \in K[X]$ such that $D = AU + BV$;*
b) *D divides A and B;*
c) *every polynomial dividing both A and B divides D.*

Consequently, D is a *greatest common divisor* of A and B. Assume that A and B are not both equal to zero. Then the ideal (A, B) is nonzero and its generators differ only by the multiplication by a nonzero element of K. In this case, we will agree to call the g.c.d. of A and B the unique monic polynomial generating (A, B). Recall that two polynomials A and B are said to be *coprime* if their only common divisors are the constant polynomials. By the preceding corollary, this amounts to saying that there exist two polynomials U and V such that $AU + BV = 1$, a statement sometimes referred to as Bézout's theorem.

Proof. I leave as an exercise to the reader the task of checking that I is actually an ideal in $K[X]$. By the very definition of D, $D \in I$ and there exist U and V in $K[X]$ such that $D = AU + BV$, hence *a*).

Since $A = A \cdot 1 + B \cdot 0$, $A \in I$ and there exists $P \in K[X]$ such that $A = PD$. Similarly, there exists $Q \in K[X]$ such that $B = QD$. It follows that A and B are both multiples of D, so that *b*) holds.

Finally, if C divides A and B, write $A = CP$ and $B = CQ$ for some polynomials P and Q. The relation $D = AU + BV$ implies $D = CPU + CQV = C(PU + QV)$, so that C divides D, which shows *c*). □

From that, one deduces that the g.c.d. of two polynomials does not depend on the field in which it is computed.

Proposition 2.4.3. *Let $K \subset L$ be a field extension, and let A and B be two polynomials in $K[X]$. Then the g.c.d. of A and B as polynomials in $L[X]$ is equal to the g.c.d. of A and B computed in $K[X]$.*

Proof. Let D be the g.c.d. of A and B in $K[X]$ and let E be their g.c.d. in $L[X]$. As D divides A and B in $K[X]$, it divides them in $L[X]$ and D divides E. To show the other divisibility, choose U and V in $K[X]$ such that $D = AU + BV$. As E divides A and B, it has to divide D! Since D and E are monic polynomials dividing each other, they are equal. □

One also deduces from Bézout's theorem the so-called Gauss's lemma, which is a crucial point in the proof that polynomial rings have the "unique factorization" property.

Lemma 2.4.4 (Gauss's lemma). *Let P be an irreducible polynomial in $K[X]$. Let A and B be two polynomials in $K[X]$ such that P divides AB. Then P divides A or P divides B.*

Proof. Assume that P does not divide A. Since P is irreducible, its only divisors are the constant polynomials $\lambda \in K^*$ and the multiples λP for $\lambda \in K^*$. Among those, only the constants divide A, so that A and P are coprime. By Bézout's theorem, we may find polynomials U and V such that $AU + PV = 1$. Multiplying this relation by B, one gets $ABU + PBV = B$. As P divides AB, one may write $AB = PQ$. Finally, $B = P(QU + BV)$ is a multiple of P, q.e.d. □

Theorem 2.4.5. *Any nonzero polynomial A in $K[X]$ admits a decomposition $A = a\prod_{i=1}^{m} P_i^{n_i}$ with $a \in K^*$, $m \geqslant 0$, P_i distinct monic irreducible polynomials, and n_i positive integers.*

Moreover, if $A = a' \prod_{j=1}^{m'} Q_j^{n'_j}$ is another decomposition, one has $a = a'$, $m = m'$ and there exists a permutation σ of $\{1, \ldots, m\}$ such that for every i, $P_i = Q_{\sigma(i)}$ and $n_i = n'_{\sigma(i)}$.

One says that the ring $K[X]$ is a *factorial ring* or a *unique factorization domain* (or ring).

Proof. The existence of such a decomposition is shown by induction on the degree of A. If A is irreducible, one just writes $A = aP$ where P is irreducible and monic and a is the leading coefficient of A. Otherwise, one may write $A = A_1A_2$ with two polynomials A_1 and A_2 whose degrees are less than $\deg A$, and we conclude by induction.

Uniqueness is the important point, and to prove that we also argue by induction. Considering the leading coefficients, we see at once that $a = a'$. The polynomial P_1 is irreducible and divides A. By Gauss's lemma, it divides one of the Q_j, say $Q_{\sigma(1)}$. Since $Q_{\sigma(1)}$ is irreducible, P_1 and $Q_{\sigma(1)}$ are multiples one of another; being monic, they are equal. Now apply the inductive hypothesis to A/P_1. □

Let us give the general definition of a factorial ring.

Definition 2.4.6. *Let A be an integral domain. One says an element a in A is* irreducible *if* a) *a is not invertible in A;* b) *for any x and y in A such that $a = xy$, either x or y is invertible in A.*

One says the ring A is factorial *if the following two properties hold:*

a) *for every nonzero element $a \in A$, there exists an integer $r \geqslant 0$, irreducible elements $p_1, \ldots, p_r$ and a unit u with $a = up_1 \ldots p_r$* (existence of a decomposition into irreducible factors)*;*

b) *if $a = up_1 \ldots p_r$ and $a = vq_1 \ldots q_s$ are two decompositions then $r = s$ and there exists a permutation σ of $\{1, \ldots, r\}$ and units u_j $(1 \leqslant j \leqslant r)$ such that for every j, $q_j = u_j p_{\sigma(j)}$* ("uniqueness" of the decomposition in irreducible factors).

In a factorial ring, any two nonzero elements have a g.c.d., which is well defined up to multiplication by a unit. The arguments of this Section show that any principal ideal ring is a factorial ring. See Exercises 2.6 and 2.7 for applications.

Theorem 2.4.7 (Gauss). *If A is a factorial ring, then $A[X]$ is a factorial ring too.*

The proof begins by describing the irreducible elements in $A[X]$; besides the irreducible elements in A, these are polynomials in $A[X]$ whose coefficients are coprime and which are irreducible as coefficients in K, where K denotes the field of fractions of A. Now we shall generalize the result proved in Exercise 1.9 to arbitrary factorial rings. Let the *content* of a nonzero polynomial in $A[X]$ be the g.c.d. of its coefficients. Then, *if P and Q are two nonzero polynomials in $A[X]$, the content of their product PQ is equal to the product of the contents of P and Q* (up to a unit).

A field is a factorial ring, and so is the ring of integers. The following important corollary follows by induction.

Corollary 2.4.8. *The rings $\mathbf{Z}[X_1, \ldots, X_n]$ and, if K is a field, $K[X_1, \ldots, X_n]$, are factorial rings.*

2.5 Appendix: Quotient rings

In this section, I explain how the construction of the ring of remainders done in Section 2.1 can be generalized.

The situation is as follows. One is given a ring A and an ideal I of A; the goal is to construct a *quotient ring*, which will be denoted A/I, and a surjective ring homomorphism $\pi\colon A \to A/I$ with kernel I. Therefore two elements a and b have the same image in A/I if and only if their difference $a-b$ belongs to I; one then says that a and b are in the same *residue class modulo I*. (*Exercise:* check that this is an equivalence relation.) In Section 2.1, we considered the case where $A = K[X]$ and $I = (P)$ is the ideal generated by a polynomial $P \in K[X]$. In that case, the remainder in the Euclidean division of a polynomial by P gives us a canonical element in the residue class modulo I of any polynomial in $K[X]$. When $A = \mathbf{Z}$ and $I = (n)$, one still has a canonical element in each class, for instance, the integers in the set $\{0, \dots, n-1\}$. This will be the case in any *Euclidean ring*, that is, in any ring admitting some kind of Euclidean division, but not in general. Such a difficulty should not bother us too much. The choice of this element has no importance at all and any element will do. A more elegant way consists of defining A/I as *the set of all residue classes modulo I* so that elements of A/I are just subsets of A; instead of choosing some element, we take them all together. If $a \in A$, let us denote by $\overline{a}$ the class of a in A/I. We define a map $\pi\colon A \to A/I$ by the simple formula $\pi(a) = \overline{a}$.

To say that A/I is a ring and that π is a ring homomorphism amounts to saying that addition and multiplication in A/I are defined to be compatible with those of A and with the map $a \mapsto \overline{a}$. One thus needs to check that if $\overline{a} = \overline{b}$ and $\overline{c} = \overline{d}$, $\overline{a+c} = \overline{b+d}$ and $\overline{ac} = \overline{bd}$, for this will allow us to define addition and multiplication in A/I by the formulae $\overline{a} + \overline{c} = \overline{a+c}$ and $\overline{a} \cdot \overline{c} = \overline{ac}$. But $(b+d) - (a+c) = (b-a) + (c-d)$ and $bd - ac = (b-a)d + a(d-c)$ are both the sum of two elements in I, so belong to I. The other axioms of a ring structure and of a ring homomorphism are checked in the same way.

If I is an ideal in A, it may be interesting to express the algebraic properties that the quotient ring A/I might possess, in terms of the ideal I.

Proposition 2.5.1. a) *The ring A/I is null if and only if $A = I$;*

b) *the ring A/I is an integral domain if and only if $I \neq A$ and if for any x and y in $A \setminus I$, $xy \notin I$;*

c) *the ring A/I is a field if and only if $I \neq A$ and if the only ideals of A containing I are I and A.*

In case *b*), one says I is a *prime ideal*; in case *c*), it is a *maximal ideal*.

Proposition 2.5.2. *Let A be a principal ideal ring which is not a field. Then its prime ideals are* a) *the null ideal* (0); b) *the ideals generated by a irreducible element.*

Among these, only the null ideal is not maximal.

The following abstract theorem concerns the existence of prime or maximal ideals in an arbitrary ring.

Theorem 2.5.3 (Krull). *Let A be a ring. Every ideal of A not equal to A is contained in a maximal ideal.*

Proof. Let I be an ideal in A, with $I \neq A$. Let us endow A with a well-ordering $\prec$.

We will define by induction increasing families $(J_x)_{x \in A}$ and $(I_x)_{x \in A}$ of ideals of A, satisfying $I \subset J_x \subset I_x \neq A$, as follows.

If x is the minimal element of A, set $J_{\prec x} = I$.

Let $x \in A$, distinct from the minimal element of A, and assume that I_y has been constructed for $y \prec x$. We first set $J_x = \bigcup_{y \prec x} I_y$; since the union is increasing, observe that J_x is an ideal of A. Indeed, let $a,\ a' \in I_{\prec x}$; there are y and $y' \prec x$ such that $a \in I_y$ and $a' \in I_{y'}$. Since the ordering $\prec$ is total, one has $y \preceq y'$ or $y' \preceq y$. In the first case, $I_y \subset I_{y'}$, hence $a + a' \in I_{y'}$; in the second case, $a + a' \in I_y$. Consequently, $a + a' \in I_{\prec x}$. Let $a \in I_{\prec x}$ and $b \in A$. If $y \prec x$ is such that $a \in I_y$, one has $ba \in I_y$, hence $ba \in I_{\prec x}$.

Since $1 \notin I_y$ for $y \prec x$, $1 \notin J_x$ and $J_x \neq A$. Moreover, J_x contains all I_y for $y \prec x$, hence J_x contains I.

Finally, consider the ideal $J_x + (x)$. If it is distinct from A, set $I_x = J_x + (x)$; otherwise, set $I_x = J_x$.

It remains to set $J = \bigcup_{y \in A} I_x$. Since the family (I_x) is increasing, this is an ideal of A, not equal to A. Moreover, for any $x \in A \setminus J$, one has $x \notin I_x$, hence $A = J_x + (x)$ by construction, and $A = J + (x)$ *a fortiori*. This shows that the ideal J is a maximal ideal of A. □

Corollary 2.5.4. *Let A be a ring. An element in A is invertible if and only if no maximal ideal contains it.*

Proof. Let $I = (a)$ be the ideal generated by the element $a \in A$. If a is invertible, there exists $b \in A$ such that $ab = 1$, so that $1 \in I$, hence $I = A$ and I cannot be contained in a maximal ideal. Consequently, there is no maximal ideal in A containing a. Conversely, if a is not a unit, $I \neq A$. By Krull's theorem 2.5.3, there is a maximal ideal containing I and this maximal ideal automatically contains a. □

2.6 Appendix: Puiseux's theorem

This appendix is devoted to Puiseux's theorem, a result which can be viewed in two different ways:

– from an analytic point of view, it shows that solutions of a polynomial equation whose coefficients are holomorphic functions (power series) can be *parametrized* and give holomorphic functions in a parameter $t^{1/n}$;

– for the algebraist, it describes explicitly the algebraic closure of the field of meromorphic functions in a neighborhood of the origin.

If $r > 0$, $\mathscr{A}(r)$ denotes the set of continuous functions on the closed disk $\overline{D}(0,r) \subset \mathbf{C}$ whose restriction to the open disc $D(0,r)$ is holomorphic. This is a ring; by the principle of isolated zeroes, it is an integral domain. For $f \in \mathscr{A}(r)$, set $\|f\| = \sup_{|z|\leqslant r} |f(z)|$. This is a norm on $\mathscr{A}(r)$, and it defines on it the topology of uniform convergence. A uniform limit of continuous functions is continuous, and a uniform limit of holomorphic functions is again holomorphic. It follows that this norm endows $\mathscr{A}(r)$ with the structure of a Banach space, and even with the structure of a *Banach algebra* since one has $\|fg\| \leqslant \|f\| \, \|g\|$ for any f and g in $\mathscr{A}(r)$.

A function f in $\mathscr{A}(r)$ has the power-series expansion

$$f(z) = \sum_{n=0}^{\infty} a_n z^n,$$

which converges for $|z| < r$, as can be seen, *e.g.*, using Cauchy estimates of derivatives of analytic functions. Two different functions have two different expansions, which will enable us to identify elements of $\mathscr{A}(r)$ to some power-series. A word on notation: we shall have to manipulate polynomials with coefficients in $\mathscr{A}(r)$, *i.e.* polynomials the coefficients of which are *functions*. We shall denote by X the polynomial indeterminate, and by z the argument of functions in $\mathscr{A}(r)$. For example, in the next theorem, $P(z^e, X)$ is the polynomial of $\mathbf{C}[X]$ obtained by evaluating each coefficient of the polynomial $P \in \mathscr{A}(r)[X]$ at z^e.

Theorem 2.6.1 (Puiseux). *Let P be a monic polynomial of degree n with coefficients in $\mathscr{A}(r)$. There exists an integer $e \geqslant 1$, a real number $\rho \in (0, r^{1/e}]$, and functions $x_1, \dots, x_n \in \mathscr{A}(\rho)$ such that*

$$P(z^e, X) = \prod_{i=1}^{n} (X - x_i(z)).$$

In particular, for $|z| < r$, the n roots of the polynomial $P(z)$ are parametrized by power series $x_i(z^{1/e})$ in a fractional power of z. Let us give some simple examples that show the necessity of introducing such a fractional power, and also that the radius of convergence of the solutions can be smaller than the one of the coefficients.

a) The roots of $P = X^2 - 2zX - 1$ are

$$x_1(z) = z + \sqrt{1+z^2} = 1 + z + \sum_{n=1}^{\infty} \binom{1/2}{n} z^{2n}$$

and

$$x_2(z) = z - \sqrt{1+z^2} = -1 + z + \sum_{n=1}^{\infty} \binom{1/2}{n} (-1)^n z^{2n},$$

two power series converging for $|z| < 1$.

b) The roots of $P = X^2 - z(1+z)$ are

$$\pm z^{1/2}\sqrt{1+z} = \pm \sum_{n=0}^{\infty} \binom{1/2}{n} (z^{1/2})^{2n+1},$$

two power series converging for $|z| < 1$. In that case, one has $e = 2$.
Theorem 2.6.1 is proved by induction on n.

Proposition 2.6.2. *Let $P \in \mathscr{A}(r)[X]$ be a monic polynomial with degree n. Let Q_0 and $R_0 \in \mathbf{C}[X]$ be two monic polynomials of degrees $< n$, such that $P(0,X) = Q_0(X)R_0(X)$. If Q_0 and R_0 are coprime, then there exists $\rho \in (0,r]$ and two monic polynomials Q and R with coefficients in $\mathscr{A}(\rho)$, such that $Q(0,X) = Q_0(X)$, $R(0,X) = R_0(X)$ and $P = QR$.*

Proof. This is an application of the implicit function theorem, in its holomorphic version. To prove it, however, we will go back to Banach's fixed-point theorem.

Set $P_0 = P(0,X)$ and let $P_1 \in \mathscr{A}(r)[X]$ be such that $P = P_0 + zP_1$. Let $m = \deg Q_0$, $p = \deg R_0$; one has $m + p = n$. We are looking for Q and R such that $Q = Q_0 + zU$ and $R = R_0 + zV$, where U has degree $< m$ and V has degree $< p$. The equation $P = QR$ can be rewritten as

$$P_1 = UR_0 + VQ_0 + zUV.$$

If a is an integer, identify $\mathbf{C}^a$ with polynomials of degree $< a$ and introduce the linear map $\varphi\colon \mathbf{C}^m \times \mathbf{C}^p \to \mathbf{C}^{m+p}$ defined by $\varphi(U,V) = UR_0 + VQ_0$. It is *injective*, for if $\varphi(U,V) = 0$, R_0 divides VQ_0 but is prime to Q_0, so it divides V. Since $\deg V < p = \deg Q_0$, that forces $V = 0$. Similarly $U = 0$. Like any injective linear map between vector spaces of the same finite dimension, φ is an isomorphism, the inverse of which, $\varphi^{-1}\colon \mathbf{C}^{m+p} \to \mathbf{C}^m \times \mathbf{C}^p$, is also linear.

Similarly, identify $\mathscr{A}(r)^a$ with polynomials of degree $< a$ with coefficients in $\mathscr{A}(r)$ and let us consider the map $\Phi\colon \mathscr{A}(r)^m \times \mathscr{A}(r)^p \times \mathscr{A}(r)^{m+p}$ given by $\Phi(U,V) = UR_0 + VQ_0$, U and V being polynomials with coefficients in $\mathscr{A}(r)$ of degrees $< m$ and $< p$. By construction, one $\Phi(U,V)(z) = \varphi(U(z),V(z))$

for any $z \in \overline{D}(0,r)$. The map Φ is bijective and its inverse is the map $\Psi\colon \mathscr{A}(r)^{m+p} \to \mathscr{A}(r)^m \times \mathscr{A}(r)^p$ defined by $\Psi(P)(z) = \varphi(P(z))$. The equation $P = QR$ can thus be rewritten as

$$(U,V) = \Psi(P_1 - zUV).$$

The right hand side of this equation will be denoted by $T(U,V)$.

For any integer a, endow $\mathscr{A}(r)^a$ with the norm $\|(f_1,\dots,f_a)\| = \|f_1\| + \dots + \|f_a\|$. Again, this a Banach space. The linear maps Φ and Ψ are continuous and Lipschitz with these norms. In fact, if $\mathbf{C}^a$ is endowed with the norm $\|(z_1,\dots,z_a)\| = |z_1| + \dots + |z_a|$, then their Lipschitz constants are the same as those of φ and φ^{-1}. Set $A = \|\Psi\|$.

For any $U \in \mathscr{A}(r)^m$ and $V \in \mathscr{A}(r)^p$, one has $\|UV\| \leqslant \|U\|\,\|V\|$. In fact, writing $U = f_0 + f_1 X + \dots + f_{m-1}X^{m-1}$ and $V = g_0 + g_1 X + \dots + g_{p-1}X^{p-1}$, one has

$$\begin{aligned}\|UV\| = \sum_{j=0}^{m+p-1} \left\| \sum_{k+\ell=j} f_k g_\ell \right\| &\leqslant \sum_{j=0}^{m+p-1} \sum_{k+\ell=j} \|f_k\|\,\|g_\ell\| \\ &\leqslant \sum_{k=0}^{m-1} \|f_k\| \sum_{\ell=0}^{p-1} \|g_\ell\| \leqslant \|U\|\,\|V\|.\end{aligned}$$

It follows that the map T from $\mathscr{A}(r)^m \times \mathscr{A}(r)^p$ to itself satisfies

$$\|T(U,V)\| \leqslant A\,\|P_1\| + Ar\,\|U\|\,\|V\|.$$

If R and r are real numbers satisfying $R > A\,\|P_1\|$, and if $r < r_1 = (R - A\,\|P_1\|)/AR^2$, then the ball B_R defined by $\|U\| + \|V\| \leqslant R$ in $\mathscr{A}(r)^{m+p}$ is stable under T.

Moreover, if (U,V) and $(U',V') \in B_R$,

$$\begin{aligned}\|T(U,V) - T(U',V')\| &= \|\Psi(-zUV + tU'V')\| \\ &\leqslant Ar\,\|UV - U'V'\| \\ &\leqslant Ar\,\|U(V-V') + V'(U-U')\| \\ &\leqslant ArR\big(\|U-U'\| + \|V-V'\|\big).\end{aligned}$$

Consequently, if $r < r_2 = 1/AR$, T is a contracting map.

It remains to observe that we can fix some $R > A\,\|P_1\|$ and then choose $\rho < \min(r, r_1, r_2)$. With those choices, the linear map T from $\mathscr{A}(\rho)^m \times \mathscr{A}(\rho)^p$ to itself stabilizes the ball B_R defined by $\|U\| + \|V\| \leqslant R$ and is contracting there. By Banach's fixed point theorem, T has a unique fixed point in B_R, hence a factorization $P = QR$ in the ring $\mathscr{A}(\rho)[X]$. □

This first step (Proposition 2.6.2) will allow us to assume that $P(0,X)$ has a unique root. Consider a factorization $P(0,X) = \prod_j (X - z_j)^{n_j}$, with *distinct* complex numbers z_j; by the Proposition, it extends to a factorization $P = \prod_j P_j$, with $P_j \in \mathscr{A}(\rho)[X]$ and $P_j(0,X) = (X - z_j)^{n_j}$. Assume that for any j, the polynomial P_j satisfies the conclusion of Puiseux's Theorem, i.e. , that there exists $e_j \geqslant 1$, and functions $x_{j,i} \in \mathscr{A}(\rho_j)$, $1 \leqslant i \leqslant n_j$, such that

$$P_j(z^{e_j}, X) = \prod_{j=1}^{n_j} (X - x_{j,i}(z)).$$

Then we may set $e = \mathrm{l.c.m.}(e_1, \dots, e_j, \dots)$ and $f = j = e/e_j$, so that

$$P(z^e, X) = \prod_j P_j\big((z^{f_j})^{e_j}, X\big) = \prod_j \prod_{i=1}^{n_j} (X - x_{j,i}(z^{f_j})),$$

which proves the assertion of Puiseux's theorem for P, with $\rho = \min(\rho_j^{1/f_j})$.

Consequently, we can assume that $P(0,X)$ has a unique root α. Replacing the polynomial $P = X^n + a_1 X^{n-1} + \dots$ by $P(X - a_1/n)$, we may moreover assume that the coefficient of X^{n-1} in P is zero, which means that the sum of all roots of P is zero. In particular, $\alpha = 0$ and $P(0,X) = X^n$.

The next proposition refers to the order of vanishing at zero of a nonzero function $f \in \mathscr{A}(r)$: if its expansion as a power series is $f = \sum_{n \geqslant 0} a_n z^n$, the order of vanishing at zero of f is the smallest integer n such that $a_n \neq 0$. It is also the highest power of z dividing f. We will denote it by $v(f)$.

Proposition 2.6.3. *Let $P = X^n + a_2 X^{n-2} + \dots + a_n$ be a monic polynomial with coefficients in $\mathscr{A}(r)$. Let $\nu = \min_{2 \leqslant j \leqslant n} v(a_j)/j$; write $\nu = m/e$ where m and e are two coprime nonnegative integers. Then there exists a monic polynomial Q, of degree n, with coefficients in $\mathscr{A}(r^{1/e})$ such that*

$$z^{mn} Q(z,X) = P(z^e, z^m X).$$

At $z = 0$, $Q(0,X) \neq X^n$.

Before we prove this proposition, let us finish the proof of Puiseux's theorem. Since $Q(0,X) \neq X^n$, and since the sum of its roots is zero, not all of the roots of $Q(0,X)$ are equal and Proposition 2.6.2 allows us to factor Q as $Q = RS$ (in a certain $\mathscr{A}(\rho)$). By induction, we thus see that there exist an integer $f \geqslant 1$, a real number ρ and power series $y_j(z) \in \mathscr{A}(\rho)$ such that

$$Q(z^f, X) = \prod_{j=1}^{n} (X - y_j(z)).$$

Thus

$$P(z^{ef}, z^m X) = z^{mn} \prod_{j=1}^{n} (X - y_j(z^f))$$

and

$$P(z^{ef}, X) = \prod_{j=1}^{n} (X - z^m y_j(z^f)),$$

so that the $x_j = z^m y_j(z^f)$ are the power series we were searching for.

Proof of Proposition 2.6.3. In the expansion

$$P(z^e, z^m X) = \sum_{j=0}^{n} a_j(z^e) z^{m(n-j)} X^{n-j},$$

the coefficient $a_j(z^e)z^{m(n-j)}$ is a power series whose order of vanishing at 0 is equal to $ev(a_j) + m(n-j) = mn + e(v(a_j) - j\nu) \geqslant mn$. Therefore one can find a power series $b_j \in \mathscr{A}(r^{1/e})$ such that $a_j(z^e)t^{m(n-j)} = z^{mn} b_j(z)$. Moreover, if the integer $j \geqslant 2$ is chosen so that $v(a_j)/j = \nu$, one has $v(b_j) = 0$, which means $b_j(0) \neq 0$. Consequently, $Q(0) \neq X^n$. □

Exercises

Exercise 2.1. **a)** If $d_1, \dots, d_r$ are positive integers, show that $d_1! \dots d_r!$ divides $(d_1 + \dots + d_r)!$.

b) Following the steps of the construction of a splitting extension for a polynomial of degree d, show that it is a finite extension and that its degree divides $d!$.

Exercise 2.2. Let p be a prime number, $p \geqslant 3$.

a) Show that $\prod_{a \in (\mathbf{Z}/p\mathbf{Z})^*} a = -1$ (*Wilson's theorem*). — Hint: in the product, group a and $1/a$, provided they are distinct.

b) For $a \in (\mathbf{Z}/p\mathbf{Z})^*$, let $S_a = \{a, -a, 1/a, -1/a\}$. Show that for a and b in $(\mathbf{Z}/p\mathbf{Z})^*$, either $S_a = S_b$, or $S_a \cap S_b = \emptyset$.

c) Computing the cardinality of S_a according to whether $a^2 = \pm 1$ or not, show that -1 is a square in $(\mathbf{Z}/p\mathbf{Z})^*$ if and only if $p \equiv 1 \pmod 4$. If it exists, can you find a formula for a square root of -1?

Exercise 2.3. An algebraically closed field is infinite.

Exercise 2.4. Let K be a field, p a prime number and let a be an element in K. Show that the polynomial $X^p - a$ is reducible in $K[X]$ if and only if it has a root in K. (If $X^p - a = P(X)Q(X)$, what can $P(0)$ be equal to?)

Exercise 2.5 (Gauss). For $n \in \mathbf{N}^*$, let $\Phi_n \in \mathbf{C}[X]$ be the monic polynomial with simple roots, given by the primitive nth roots of unity in $\mathbf{C}$.

a) Show that $\prod_{d|n} \Phi_d = X^n - 1$. Deduce by induction that for any n, $\Phi_n \in \mathbf{Z}[X]$.

Let ζ be any primitive nth root of unity and let $P \in \mathbf{Q}[X]$ be its monic minimal polynomial.

b) Show that P has integer coefficients and that it divides Φ_n.

c) Let p be a prime number. Show that there exists a polynomial $Q \in \mathbf{Z}[X]$ such that $P(X^p) - P(X)^p = pQ(X)$. Prove the existence of $b \in \mathbf{Z}[\zeta]$ such that $P(\zeta^p) = pb$.

d) Let p be a prime number that does not divide n. If $P(\zeta^p) \neq 0$, show by differentiating the polynomial $X^n - 1$ that there exists $c \in \mathbf{Z}[\zeta]$ with $n\zeta^{n-1} = pc$. Deduce from that a contradiction, hence $P(\zeta^p) = 0$.

e) Show that $P = \Phi_n$, that is the polynomial Φ_n is irreducible in $\mathbf{Q}[X]$.

Exercise 2.6. Let A be the subring $\mathbf{Z}[i]$ in $\mathbf{C}$ (ring of Gaussian integers).

a) Show that for any a and b in A, with $b \neq 0$, there exist q and r in A with $a = bq + r$ and $|r| < |b|$.

b) Show that A is a principal ideal ring. In particular, it is a factorial ring.

c) Let p be a prime number. Show that one of the following is true: 1) either p is irreducible in A; or 2) there exist a and b in $\mathbf{N}$ such that $p = a^2 + b^2$, and $p = (a + ib)(a - ib)$ is a decomposition of p as a product of irreducible elements in A.

d) Show that prime numbers congruent to 3 modulo 4 are irreducible in A. Show that 2 is not.

e) Let p be a prime number. Define a ring isomorphism from A/pA to the ring $(\mathbf{Z}/p\mathbf{Z})[X]/(X^2 + 1)$. Deduce that p is reducible in A if and only if the polynomial $X^2 + 1$ has a root in the field $\mathbf{Z}/p\mathbf{Z}$. By Exercise 2.2, this happens exactly when p is equal to 1 modulo 4.

In particular, prime numbers equal to 1 modulo 4 are sums of two squares of integers (Fermat, 1659).

Exercise 2.7 (Every integer is a sum of four squares). Let $\mathbf{H}$ be the noncommutative field of quaternions. We identify it with $\mathbf{Q}^4$, the canonical basis of which is denoted $(1, i, j, k)$ and with multiplication defined by $i^2 = j^2 = k^2 = -1$ and $ij = k$.

a) If $z = a + bi + cj + dk \in \mathbf{H}$, set $\overline{z} = a - bi - cj - dk$ and $N(z) = z\overline{z}$. Show that $N(z) = a^2 + b^2 + c^2 + d^2$ and that $N(zz') = N(z)N(z')$. Conclude that if two integers are sums of four squares of integers, then their product is again a sum of four squares.

b) Show that the set A_0 of $x + yi + zj + tk \in \mathbf{H}$ with $x, y, z, t \in \mathbf{Z}$ is a (noncommutative) subring of $\mathbf{H}$.

c) Let $\varepsilon = (1 + i + j + k)/2$. Compute ε^2. Show that the set A of all $a \in \mathbf{H}$ such that either $a \in A_0$ or $a - \varepsilon \in A_0$ is a subring of $\mathbf{H}$.

If $z \in A$, show that $N(z) \in \mathbf{N}$. (This is clear for $z \in A_0$. Otherwise, find $u = \frac{1}{2}(\pm 1 \pm i \pm j \pm k) \in A^*$ and $b \in A_0$ with $z = u + 2b$. Observe that zu^{-1} belongs to A_0.) Show also that $z \in A$ is invertible if and only if $N(z) = 1$.

d) Show that A is a Euclidean ring: if a and $b \in A$, with $b \neq 0$, find q and $r \in A$ with $N(r) < N(b)$ and $a = bq + r$. Deduce that any (left) ideal in A is a principal ideal (of the form Az for some $z \in A$).

e) Let p be an odd prime number. Show that there exist integers a and b such that $a^2 + b^2 + 1$ is divisible by p. (How many elements of $\mathbf{Z}/p\mathbf{Z}$ are of the form $x^2 + 1$? and of the form $-y^2$?) Let I be the left ideal in A generated by p and $1 + ai + bj$. If $I = Az$, show that $N(z) = p$ and conclude that p is a sum of four squares of integers.

f) Show that for any integer $n \geqslant 0$, there exist integers a, b, c and d such that $n = a^2 + b^2 + c^2 + d^2$; any positive integer is a sum of four squares of integers (Lagrange, 1770).

Exercise 2.8. Prove that the only ideals of a field are itself and the null ideal. Conversely, show that a nonzero ring admitting only these two ideals is a field.

Exercise 2.9. Let A be the subring $\mathbf{Z}[\sqrt{-5}]$ in $\mathbf{C}$.

a) Show that any element of A can be written in a unique way as $a + b\sqrt{-5}$ with integers a and b. Show that the map $N \colon A \to \mathbf{Z}$ defined by $N(a+b\sqrt{-5}) = a^2 + 5b^2$ satisfies $N(xy) = N(x)N(y)$.

b) Show that an element $x \in A$ is a unit if and only if $N(x) = 1$.

c) Show that the elements 2, 3, $1 + \sqrt{-5}$ et $1 - \sqrt{-5}$ are irreducible in A.

d) Conclude that A is not a factorial ring.

Exercise 2.10. Let A be a ring.

a) Let I and J be two ideals of A. Show that the set $I + J$ consisting of sums $a + b$ with $a \in I$ and $b \in J$ is an ideal of A.

b) Let I be an ideal in A. Let R_I be the set of $a \in A$ such that there exists $n \in \mathbf{N}$ with $a^n \in I$. Show that R_I is an ideal in A, and that it contains I. If $I \neq A$, show that $R_I \neq A$.

c) If $A = \mathbf{Z}$, $I = (12)$, compute R_I. Generalize to any principal ideal ring.

Exercise 2.11. Let K be a field.

a) Show that the two polynomials X and Y in $K[X, Y]$ are coprime.

b) Let $I = (X, Y)$ be the ideal in $K[X, Y]$ that they generate. Show that for any polynomial $P \in I$, one has $P(0, 0) = 0$. Conclude that there is no U and V in $K[X, Y]$ such that $UX + VY = 1$.

c) Show that the map $A \to K$, $P \mapsto P(0, 0)$ is a ring morphism, with kernel I. Show that I is a maximal ideal in $K[X, Y]$.

Exercise 2.12. One says that a ring A is a *Noetherian ring* if any ideal in A is generated by a finite number of its elements.

a) If K is a field, show that $K[X]$ is Noetherian.

b) If A is Noetherian and if I is an ideal in A, show that the quotient ring A/I is Noetherian too.

c) Show that a ring is Noetherian if and only if any increasing sequence of ideals is ultimately constant.

Exercise 2.13 (Hilbert's theorem). Let A be a Noetherian ring and let $B = A[X]$. This exercise aims to prove that B is also a Noetherian ring. Let I be an ideal in $A[X]$.

For any integer n, let J_n be the ideal in A generated by the leading coefficients of polynomials $P \in I$ which have degree n.

a) Show that for any n, $J_n \subset J_{n+1}$. Deduce that there exists an integer N with $J_n = J_N$ for $n \geqslant N$.

b) For any integer n, show that there exist polynomials $P_{n,1}, \dots, P_{n,m_n} \in I$ of degree n whose leading coefficients generate J_n.

c) Show that the polynomials $P_{n,j}$ for $n \leqslant N$ and $1 \leqslant j \leqslant m_n$ generate I. (Proceed by induction on the degree: if I_0 denotes the ideal of B generated by these polynomials, and if $P \in I$ has degree n, construct a polynomial $P_n \in I_0$ such that $P - P_0$ has degree $\leqslant n - 1$.)

d) For any field K, show that $K[X_1, \dots, X_n]$ is a Noetherian ring. Similarly, show that $\mathbf{Z}[X_1, \dots, X_n]$ is a Noetherian ring.

Exercise 2.14. In a factorial ring, irreducible elements generate prime ideals.

3

Galois theory

In this chapter we establish Galois correspondence. Discovered in 1832, it describes all subextensions of the spliting extension of a (separable) polynomial in terms of a subgroup of the group of permutations of the roots of this polynomial.

This chapter is really the heart of this book. Later we shall see numerous applications. For example, Galois correspondence is the key to the problem of solvability with radicals, and to the constructions with ruler and compass.

3.1 Homomorphisms of an extension in an algebraic closure

In this section, we study the following situation: Let $j\colon K \to L$ be a finite algebraic extension and let $i\colon K \to \Omega$ be an algebraic closure of K. Can we find morphisms of extensions from L to Ω, that is, field homomorphisms $f\colon L \to \Omega$ such that $f \circ j = i$? In other words, is it possible to *extend* the morphism i from K to L? Such an f is called a K*-homomorphism* from L to Ω. We already studied an instance of this problem in proving Theorem 2.1.5, when L is of the form $K[X]/(P)$ for an irreducible polynomial P.

Definition 3.1.1. *One says an extension $j\colon K \to L$ is* simple *if there exists $x \in L$ such that $L = K[x]$.*

Proposition 3.1.2 (Corollary of Theorem 2.1.5). *Let $j\colon K \to L$ be a simple extension and let $x \in L$ such that $L = K[x]$ with P its minimal polynomial. Let $i\colon K \to \Omega$ be an algebraic closure of K. There is a bijection between the set of K-homomorphisms from L to Ω and the set of roots of P in Ω, given by the map $f \mapsto f(x)$. In particular, there is at least one of these morphisms, and at most $[L : K]$.*

Remark 3.1.3. Each of these homomorphisms allows us to consider Ω as an algebraic closure of L. But these morphisms are all different, and for this reason it is better to study field extensions as (injective) morphisms rather than subfields. However, once such a morphism is *fixed*, there is often no harm in replacing L by its image in Ω, which puts us in the possibly more reassuring situation of subfields $K \subset L \subset \Omega$.

A polynomial $P \in K[X]$ is *separable* if its roots in an algebraic closure of K are simple.

Lemma 3.1.4. *Let K be a field. A polynomial $P \in K[X]$ is separable if and only if P and P' are coprime.*

Proof. Let Ω be an algebraic closure of P. By definition, P is separable if and only if P and P' are coprime in $\Omega[X]$. By Corollary 2.4.3, this is equivalent to the fact that P and P' are coprime in $K[X]$. □

If $K \to L$ is an algebraic extension, we shall say an element $\alpha \in L$ is *separable* over K if its minimal polynomial is separable.

Lemma 3.1.5. *Let $K \to L$ be an algebraic extension and let Ω be an algebraic closure of L. If $\alpha \in \Omega$ is separable over K, it is also separable over L.*

Note: under these conditions, α is algebraic over L and the extension $K \to L$ is algebraic, so that α is algebraic over K (Theorem 1.3.16).

Proof. Let P be the minimal polynomial of α over L and let Q be its minimal polynomial over K. Since $Q(\alpha) = 0$, Q is multiple of P. Since α is assumed to be separable over K, Q has simple roots in Ω, and therefore so does P. □

Theorem 3.1.6. *Let K be a field, $j\colon K \to L$ a finite extension and $i\colon K \to \Omega$ an algebraic closure. Then the number N of distinct K-homomorphisms from L to Ω satisfies $1 \leqslant N \leqslant [L : K]$. Moreover, the following three properties are equivalent:*

a) $N = [L : K]$;

b) *there are elements $x_1, \dots, x_n \in L$, separable over K, such that $L = K[x_1, \dots, x_n]$;*

c) *any element in L is separable over K.*

We will say that an extension $K \to L$ satisfying these properties is *separable.*

Proof. Since L is a finite extension of K, there are elements $x_1, \dots, x_n \in L$ such that $L = K[x_1, \dots, x_n]$. The proof now follows by induction on n. For $n = 1$, L is a simple extension and, by Proposition 3.1.2, N is equal to the number of distinct roots in Ω of the minimal polynomial of x_1. Since this polynomial has degree $[L : K]$, one knows these two facts:

- the integer N lies between 1 and $[L : K]$;
- equality $N = [L : K]$ holds if and only if x_1 is separable over K.

Assume that $x_1 \notin K$; let P_1 be its minimal polynomial, d its degree and set $L_1 = K[x_1]$. The restriction to L_1 of any K-morphism $f\colon L \to \Omega$ is a K-homomorphism f_1 from L_1 to Ω, hence corresponds to the choice of a root of P_1 in Ω. Therefore, there are between 1 and d such homomorphisms f_1, and each of them allows us to view Ω as an algebraic closure of L_1. The corresponding situation is summed up by the following diagram.

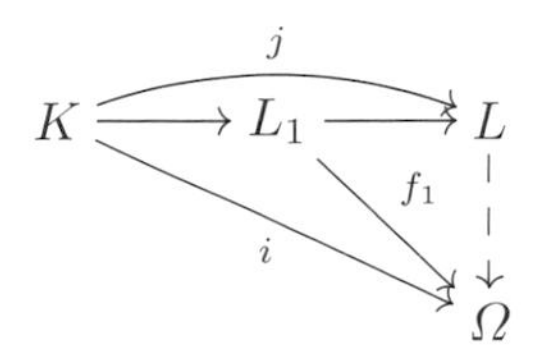

The degree of the extension $L_1 \subset L$ is equal to $[L : K]/d$, hence is less than $[L : K]$. By induction, for any homomorphism f_1, the number of L_1-homomorphisms from L to Ω which extend it lies between 1 and $[L : L_1]$. In this way, we have constructed distinct K-homomorphisms from L to Ω, in particular between 1 and $[L : K]$ of them. Since any K-homomorphism from L to Ω can be obtained via this procedure, this proves the first part of the theorem.

The preceding proof also shows that the equality $N = [L : K]$ holds if and only if x_1 is separable over K, x_2 is separable over $K[x_1]$, etc. By Lemma 3.1.5, this condition is satisfied when all of the x_i are separable over K, hence $N = [L : K]$ in that case. Assume now that $N = [L : K]$. For any $x \in L$, the previous argument applied to the family $(x, x_1, \dots, x_n)$ shows that x is separable over K. This shows that L is separable over K. □

Definition 3.1.7. *One says a field K is* perfect *if any irreducible polynomial in $K[X]$ has as many distinct roots in an algebraic closure as its degree.*

By the very definition of a perfect field, Theorem 3.1.6 implies that the following properties are equivalent:

a) K is a perfect field;

b) any irreducible polynomial of $K[X]$ is separable;

c) any element of an algebraic closure of K is separable over K;

d) any algebraic extension of K is separable;

e) for any finite extension $K \to L$, the number of K-homomorphisms from L to an algebraically closed extension of K is equal to $[L : K]$.

Corollary 3.1.8. *Any algebraic extension of a perfect field is again a perfect field.*

Proof. This is a reformulation of Lemma 3.1.5. Let K be a perfect field and let $K \to L$ be a finite extension of K. An element which is algebraic over L is algebraic over K. If it is separable over K, it is also separable over L by the lemma. □

We conclude this section with a characterization of perfect fields that does not involve extensions.

Proposition 3.1.9. *The perfect fields are* a) *fields of characteristic zero;* b) *fields of characteristic* $p > 0$ *whose Frobenius homomorphism is bijective.*

In particular, *finite fields are perfect* : in that case, the Frobenius homomorphism, as any injective self-map of a finite set, is automatically bijective.

Proof. Let P be any monic irreducible polynomial in $K[X]$. Multiple roots of P in an algebraic closure Ω are exactly the roots common to P and P', that is, the roots of their g.c.d. D. Since P is irreducible, this g.c.d. is either equal to 1 or to P. If $D = 1$, all roots are simple. If $D = P$, all are multiple and in that case P divides P'. But, the degree of P' being at most $\deg P - 1$, this implies $P' = 0$.

In characteristic zero, this is of course impossible: if the leading monomial of P is aX^n, with $n = \deg P$, the leading term of P' is naX^{n-1}, and since $na \neq 0$, $\deg P' = n - 1$. All fields of characteristic zero are thus perfect.

If however K has characteristic $p > 0$, the equality $P' = 0$ means that the degree of any nonzero term in P is divisible by p. Consequently, $P = a_n X^{pn} + a_{n-1} X^{p(n-1)} + \cdots + a_0$ is a polynomial in X^p. Assume that the Frobenius morphism of K is surjective. For any n, there exists $b_n \in K$ such that $b_n^p = a_n$. Then

$$P = b_n^p X^{pn} + b_{n-1}^p X^{p(n-1)} + \cdots + b_0^p = (b_n X^n + \cdots + b_0)^p ,$$

which contradicts the irreducibility of P. Such a field is therefore perfect. Conversely, if the Frobenius homomorphism of K is not surjective, choose $a \in K$ which is not a pth power and let $P = X^p - a$. In Ω, a has a pth power, say b, so that $P = (X - b)^p$ in $\Omega[X]$ and P has a root with multiplicity p. However, P is irreducible. Assume that $P = QR$ for two nonconstant monic polynomials Q and $R \in K[X]$. It follows that $Q = (X - b)^m$ and $R = (X - b)^{p-m}$ in $\Omega[X]$, for some integer m with $1 \leqslant m < p$. By expanding Q, we find that $mb \in K$ and since $m \neq 0$ in K, $b \in K$, which contradicts the hypothesis that a is not a pth power in K. This shows that such fields are not perfect, and concludes the proof of the proposition. □

3.2 Automorphism group of an extension

Definition 3.2.1. *Let* $j\colon K \to L$ *be a field extension. A* K-automorphism of L *is field automorphism* $\sigma\colon L \to L$ *which is a morphism of extensions.*

The set of all K-automorphisms of L is a group (under composition), denoted $\mathrm{Aut}(L/K)$. Note an important particular case: if $j\colon K \to L$ is the inclusion of a subfield $K \subset L$, a K-automorphism of L is a field automorphism $\sigma\colon L \to L$ which restricts to the identity on K.

Interest in this notion comes from the following obvious, but crucial, remark. Let σ be a K-automorphism of L and let $P \in K[X]$. For any $x \in L$, one has $\sigma(P(x)) = P(\sigma(x))$. In particular, if x is a root of P, then so is $\sigma(x)$. This means that σ *permutes* the roots of P in L.

Examples 3.2.2. *a*) Consider the extension $\mathbf{R} \subset \mathbf{C}$. Let σ be a $\mathbf{R}$-automorphism of $\mathbf{C}$. If $z = a + ib \in \mathbf{C}$, with $a, b \in \mathbf{R}$,

$$\sigma(z) = \sigma(a+ib) = \sigma(a) + \sigma(ib) = a + \sigma(i)b.$$

Since

$$\sigma(i)^2 = \sigma(i^2) = \sigma(-1) = -1,$$

one has $\sigma(i) = \pm i$, hence two automorphisms: the identity and complex conjugation.

b) Let ω denote the real number $\sqrt[3]{2}$ and let us consider the extension $\mathbf{Q} \subset \mathbf{Q}(\omega)$. Since any element in $\mathbf{Q}(\omega)$ can be written $a + b\omega + c\omega^2$, for some $a, b, c \in \mathbf{Q}$, a $\mathbf{Q}$-automorphism σ of $\mathbf{Q}(\omega)$ is well defined once the image of ω is given. But

$$\sigma(\omega)^3 = \sigma(\omega^3) = \sigma(2) = 2,$$

and the equation $x^3 = 2$ has only one real root, namely ω, so ω is its only root in $\mathbf{Q}(\omega)$. Consequently, $\mathrm{Aut}(\mathbf{Q}(\omega)/\mathbf{Q}) = \{\mathrm{id}\}$.

c) One has $\mathrm{Aut}(\mathbf{R}/\mathbf{Q}) = \{\mathrm{id}\}$ (*exercise*!).

d) Any $\mathbf{C}$-automorphism of $\mathbf{C}(X)$ is given by $P \mapsto P((aX+b)/(cX+d))$ for some matrix $\left(\begin{smallmatrix} a & b \\ c & d \end{smallmatrix}\right)$ in $\mathrm{GL}(2, \mathbf{C})$, well defined up to multiplication by a nonzero scalar. It follows that $\mathrm{Aut}(\mathbf{C}(X)/\mathbf{C}) = \mathrm{PGL}(2, \mathbf{C})$. (See Exercise 3.4.)

Remark 3.2.3. Let $K \subset L$ be a finite extension and let $\sigma\colon L \to L$ by any K-homomorphism. Like any morphism of fields, σ is injective. It follows that $\sigma(L)$ is a K-vector space of dimension $[L : K]$, hence $[\sigma(L) : K] = [L : K]$. Since $\sigma(L) \subset L$, it follows that $\sigma(L) = L$ and σ is surjective. Consequently, σ is a K-automorphism.

Proposition 3.2.4. *Let* $j\colon K \to L$ *be a finite extension. The cardinality of* $\mathrm{Aut}(L/K)$ *is at most* $[L : K]$. *If the bound is attained, then the extension* $K \to L$ *is separable.*

Proof. Let $i\colon L \to \Omega$ be some algebraic closure of L. Any K-automorphism $\sigma \in \operatorname{Aut}(L/K)$ determines a K-homomorphism $i\circ\sigma\colon L \to \Omega$, and two distinct K-automorphisms of L define distinct homomorphisms from L to Ω. By Theorem 3.1.6, the number of such homomorphisms is less or equal than $[L : K]$ and if it is equal, the extension is separable. □

Definition 3.2.5. *One says a finite extension $K \to L$ is a* Galois extension, *or is* Galois, *if* $\operatorname{Aut}(L/K)$ *has cardinality* $[L : K]$. *The group* $\operatorname{Aut}(L/K)$ *is then called the* Galois group *of this extension, and is denoted* $\operatorname{Gal}(L/K)$.

By Proposition 3.2.4, a Galois extension is necessarily separable.

Let us now state the *main theorem* of Galois theory.

Theorem 3.2.6 (Galois correspondence). *Let $K \subset L$ be a finite Galois extension with Galois group $G = \operatorname{Gal}(L/K)$.*

a) *For any subgroup $H \subset G$, the set $L^H = \{x \in L\,;\ \forall \sigma \in H,\ \sigma(x) = x\}$ is a subfield of L containing K. Moreover, $[L^H : K]$ is equal to the index $(G : H)$ of H in G.*

b) *For any field E with $K \subset E \subset L$, the extension $E \subset L$ is Galois and its Galois group is $\operatorname{Gal}(L/E) = \{\sigma \in \operatorname{Gal}(L/K)\,;\ \forall x \in E,\ \sigma(x) = x\}$.*

c) *The maps $H \mapsto L^H$ and $E \mapsto \operatorname{Gal}(L/E)$ are decreasing bijections between the set of subgroups in G and the set of subfields in L that contain K, one bijection being the inverse of the other.*

The proof of this theorem needs two other statements.

Proposition 3.2.7. *Let $K \subset L$ be a finite extension. The following conditions are equivalent:*

a) *the extension $K \subset L$ is Galois;*

b) *the extension $K \subset L$ is separable and any K-homomorphism from L to an algebraic closure of L has image L;*

c) *the extension $K \subset L$ is separable and any irreducible polynomial in $K[X]$ having a root in L is already split in L;*

d) *there exists a separable polynomial $P \in K[X]$ of which the extension $K \subset L$ is a splitting extension.*

Proof. Let $i\colon L \to \Omega$ be an algebraic closure of L, fixed for the rest of the proof.

Assume *a)*. Any element $\sigma \in \operatorname{Gal}(L/K)$ defines a K-homomorphism from L to Ω, namely $i \circ \sigma$. Since there are at most $[L : K]$ of these morphisms, they are all defined in this way, hence *b)*.

Assume that the extension $K \subset L$ is separable. Then, Theorem 3.1.6 asserts that there are $[L : K]$ morphisms of extensions from L to Ω. If *b)* holds, their image is L, so that they define distinct K-automorphisms of L.

It follows that $\operatorname{Aut}(L/K)$ has cardinality at least $[L : K]$, and the other inequality being always true, the extension $K \subset L$ is Galois.

Still assuming *b)*, let P be an irreducible polynomial in $K[X]$ which has a root ω in L. Let $E = K[\omega] \subset L$ denote the subfield of L generated by this root. For any root $\alpha \in \Omega$ of P, there is a unique K-homomorphism from E to Ω such that $\omega \mapsto \alpha$. By Theorem 3.1.6 applied to the extension $E \subset L$, such a homomorphism then extends to a K-homomorphism $\sigma\colon L \to \Omega$ satisfying $\sigma(\omega) = \alpha$. Since, by hypothesis, $\sigma(L) = L$, one has $\alpha \in L$ and P is split in L.

Now assume that *c)* is satisfied and let $x_1, \dots, x_n$ be elements of L such that $L = K[x_1, \dots, x_n]$. The minimal polynomial $P_i \in K[X]$ of x_i is irreducible and has a root in L. It follows that they are split in L and since the extension $K \subset L$ is separable, they have simple roots. Their least common multiple $\operatorname{l.c.m.}(P_1, \dots, P_n)$ is then split with simple roots in L, so that $K \subset L$ is a splitting extension of the separable polynomial P.

Assume finally that L is a splitting extension of a separable polynomial P and let us show that the extension $K \subset L$ is Galois. It is enough to show that any K-homomorphism σ from L to Ω maps onto L. If $x_1, \dots, x_n$ are the roots of P in L, their images $\sigma(x_1), \dots, \sigma(x_n)$ are again roots of P, so belong to L. As $L = K[x_1, \dots, x_n]$, $\sigma(L) \subset L$. By Remark 3.2.3, $\sigma \in \operatorname{Aut}(L/K)$. □

Proposition 3.2.8 (Artin's lemma). *Let L be field and let G be a finite group of automorphisms of L. Let $K = L^G$ the set of $x \in L$ such that for every $\sigma \in G$ $\sigma(x) = x$. Then K is a subfield of L such that $[L : K] = \operatorname{card} G$. In particular, the extension $K \subset L$ is Galois with group G.*

Emil Artin (1898–1962)

Proof. Let uscheck quickly that K is a subfield of L. If $\sigma \in G$, $\sigma(0) = 0$ and $\sigma(1) = 1$, hence 0 and 1 belong to K. The relations $\sigma(x+y) = \sigma(x) + \sigma(y) = x + y$ and $\sigma(xy) = \sigma(x)\sigma(y) = xy$, for x and $y \in K$, and $\sigma \in G$, show that $x+y$ and xy belong to K. Finally, if $x \in K$ and $\sigma \in G$, $\sigma(-x) = -\sigma(x) = -x$ and $-x \in K$. If moreover, $x \neq 0$, $\sigma(1/x) = \sigma(1)/\sigma(x) = 1/x$ and $1/x \in K$. It follows that K is actually a subfield of L.

If we assume that $[L : K] > \operatorname{card} G$, we can find $n = 1 + \operatorname{card} G$ elements in L, say $a_1, \dots, a_n$, which are linearly independent over K. Since $n > \operatorname{card} G$, the system of $\operatorname{card} G$ linear equations with n unknowns and coefficients in L,

$$\sum_{j=1}^{n} \sigma(a_i)x_i = 0, \qquad \sigma \in G,$$

has nonzero solutions $(x_1, \dots, x_n)$ in L^n. Take one of them with the smallest possible number of nonzero coefficients. Up to reordering the a_i's, we may assume that $x_1, \dots, x_m$ are nonzero, and that all other are 0. By linearity, we may also assume $x_m = 1$, hence the relations

$$\sum_{j=1}^{m-1} \sigma(a_i)x_i + \sigma(a_m) = 0, \qquad \sigma \in G.$$

Let $\tau \in G$ and apply τ to the preceding relation for $\tau^{-1} \circ \sigma$. One gets

$$\sum_{j=1}^{m-1} \sigma(a_i)\tau(x_i) + \sigma(a_m) = 0, \qquad \sigma \in G,$$

hence, if we substract the relation for σ,

$$\sum_{j=1}^{m-1} \sigma(a_i)(\tau(x_i) - x_i) = 0, \qquad \sigma \in G.$$

By minimality of m, one necessarily has $\tau(x_i) = x_i$ for every i and every τ. It follows that $x_i \in K$ and and the relation $\sum_{j=1}^{m} a_j x_j = 0$ is now a nontrivial dependence relation over K, although the x_i were assumed to be linearly independent.

One thus has $[L : K] \leqslant \operatorname{card} G$; in particular, the extension $K \subset L$ is finite. The elements of G can clearly be identified to K-automorphisms of L, hence an inclusion $G \subset \operatorname{Aut}(L/K)$. By Proposition 3.2.4, this implies that $\operatorname{card} G \leqslant [L : K]$, hence the equality $\operatorname{card} G = [L : K]$. This shows that the extension $K \subset L$ is Galois and its Galois group, $\operatorname{Gal}(L/K) = \operatorname{Aut}(L/K)$ identifies with G. □

We now can prove Theorem 3.2.6 (Galois correspondence). Let $K \subset L$ be a finite Galois extension with Galois group G.

Let H be a subgroup of G. By Proposition 3.2.8, the extension $L^H \subset L$ is Galois with Galois group H. It is obvious that $K \subset L^H$. Moreover,

$$[L^H : K] = \frac{[L : K]}{[L : L^H]} = \frac{\operatorname{card} G}{\operatorname{card} H} = (G : H).$$

Conversely, let E be a subfield of L, with $K \subset E$. Since the extension $K \subset L$ is Galois, we know from Proposition 3.2.7 that it is a splitting extension of some separable polynomial $P \in K[X]$. Now, $E \subset L$ is also a splitting extension of this polynomial P, now viewed as a polynomial in $E[X]$. By Proposition 3.2.7 again, the extension $E \subset L$ is Galois. Since the E-automorphisms of L are precisely the K-automorphisms of L whose restriction to E is the identity,

$$\operatorname{Gal}(L/E) = \{\sigma \in G\,;\, \forall x \in E,\ \sigma(x) = x\}.$$

Denoting by H this group, one has in particular $\operatorname{card} H = [L : E]$ and

$$(G : H) = \frac{\operatorname{card} G}{\operatorname{card} H} = \frac{[L : K]}{[L : E]} = [E : K].$$

By the first part of Theorem 3.2.6, or directly by Proposition 3.2.8, the extension $L^H \subset L$ is Galois with group H. Consequently, $[L : L^H] = \operatorname{card} H = [L : E]$; since L^H contains E, these two fields are equal.

This shows that the maps $H \mapsto L^H$ and $E \mapsto \operatorname{Gal}(L/E)$ are inverses of one another.

The word "decreasing" just means the two obvious facts: *a*) if $H \subset H'$, then $L^{H'} \subset L^H$, and *b*) if $E \subset E'$, then $\operatorname{Gal}(L/E')$ is a subgroup of $\operatorname{Gal}(L/E)$.

Proposition 3.2.9. *Let $K \subset L$ be a Galois extension with group $G = \operatorname{Gal}(L/K)$. Let H be a subgroup of G.*

a) *If $\sigma \in \operatorname{Gal}(L/K)$, one has $\sigma(L^H) = L^{\sigma H \sigma^{-1}}$.*

Let $N_G(H) = \{\sigma \in \operatorname{Gal}(L/K)\,;\, \sigma H \sigma^{-1} = H\}$, the normalizer of H in $\operatorname{Gal}(L/K)$. Elements of $N_G(H)$ are those $\sigma \in \operatorname{Gal}(L/K)$ mapping L^H to itself.

b) *Restricting an element of $N_G(H)$ to L^H defines a surjective group morphism $N_G(H)$ to $\operatorname{Aut}(L^H/K)$ with kernel H. In particular, the extension $K \subset L^H$ is Galois if and only if H is a normal subgroup in G. Then $\operatorname{Gal}(L^H/K) = G/H$.*

Proof. *a*) An element $x \in L$ belongs to L^H if and only if $h(x) = x$ for every $h \in H$. Therefore y belongs to $\sigma(L^H)$ if and only if $h\sigma^{-1}(y) = \sigma^{-1}(y)$ for any $h \in H$, that is $\sigma h \sigma^{-1}(y) = y$ for any $h \in H$, that is $y \in L^{\sigma H \sigma^{-1}}$.

b) Since the extension $K \subset L$ is Galois, any K-homomorphism $L^H \to L^H$ is the restriction to L^H of some K-homomorphism $L \to L$, that is, of an element $\sigma \in \operatorname{Gal}(L/K)$. Such a σ satisfies $\sigma(L^H) = L^H$ if and only if $\sigma H \sigma^{-1} = H$, hence a surjective group morphism $N_G(H) \to \operatorname{Aut}(L^H/K)$, $\sigma \mapsto \sigma|_{L^H}$. The kernel of this morphism is the set of all $\sigma \in N_G(H)$ such that $\sigma(x) = x$ for any $x \in L^H$, so is equal to H. It follows that $\operatorname{card} \operatorname{Aut}(L^H/K) = \operatorname{card} N_G(H)/\operatorname{card} H = (N_G(H) : H)$. (We may also take the quotient and obtain an isomorphism $N_G(H)/H \simeq \operatorname{Aut}(L^H/K)$.)

Recall that $[L^H : K] = (G : H)$. It follows that $\operatorname{card} \operatorname{Aut}(L^H/K) = [L^H : K]$ if and only if $N_G(H) = G$, which means precisely that H is a normal subgroup of G (see Definition 4.4.1). Consequently, $K \subset L^H$ is a Galois extension if and only if H is normal in G. □

Galois theory implies useful and important results, even for non-Galois extensions. This is due to the fact, which we prove now as a corollary of Prop. 3.2.7, that any finite separable extension can be embedded in a Galois extension.

Évariste Galois (1811–1832)

Proposition 3.2.10. *Let K be a field, fix an algebraic closure Ω of K and let L be a finite separable extension of K contained in Ω. In this situation, there is a smallest finite extension $L \subset L^{\mathrm{g}}$ contained in Ω such that the extension $K \subset L^{\mathrm{g}}$ is Galois.*

Proof. There are elements $x_1, \ldots, x_n$ in L such that $L = K[x_1, \ldots, x_n]$. They are separable over K, hence for any i, the minimal polynomial P_i of x_i is separable. Define L^{g} as the subfield of Ω generated by the roots of the P_i in Ω. This is a splitting extension of the separable polynomial $\mathrm{l.c.m.}(P_1, \ldots, P_n)$, hence it is a Galois extension. Moreover, for any finite extension E of L such that the extension $K \subset L$ is Galois, the polynomials P_i are split in E (see Prop. 3.2.7, c)), so that $L^{\mathrm{g}} \subset E$. The proposition is thus proved. □

Under the hypotheses of the proposition, any subfield E of L^{g} containing K corresponds to a subgroup of $\mathrm{Gal}(L^{\mathrm{g}}/K)$, namely $\mathrm{Gal}(L^{\mathrm{g}}/E)$, and the field L corresponds to the subgroup $\mathrm{Gal}(L^{\mathrm{g}}/L)$. Consequently, the set of subfields E with $K \subset E \subset L$ is in bijection with the set of subgroups of $\mathrm{Gal}(L^{\mathrm{g}}/K)$ which contain $\mathrm{Gal}(L^{\mathrm{g}}/L)$. Since a finite group has only finitely many subgroups, one deduces the following corollary.

Corollary 3.2.11. *Let $K \subset L$ be a finite separable extension. Then, there are only finitely many fields E with $K \subset E \subset L$.*

This result might look surprising and may even be *false* if the extension $K \subset L$ is not assumed to be separable; see Exercise 3.10.

3.3 The Galois group as a permutation group

In the previous section, we concentrated on fixed, "abstract" Galois extensions. In particuler, we proved that they are splitting extensions of some polynomial. But concrete problems of field theory are more likely to require another point of view: given a polynomial (irreducible, for example), what can be said of the field generated by its roots in some algebraically closed extension?

The following lemma is obvious, but one always has to keep it in mind. It claims that if $K \subset L$ is the splitting extension of a separable polynomial $P \in K[X]$, then $\mathrm{Gal}(L/K)$ permutes the roots of P and an element in $\mathrm{Gal}(L/K)$ is determined by its action on these roots.

Lemma 3.3.1. *Let K be a field, $P \in K[X]$ a separable polynomial and $K \subset L$ a splitting extension of P. Let $\mathscr{R} \subset L$ denote the set of roots of P in L.*

For any K-automorphism $\sigma \in \mathrm{Gal}(L/K)$ and any root $x \in \mathscr{R}$, $\sigma(x) \in \mathscr{R}$. The restriction of σ to $\mathscr{R}$ induces a permutation of $\mathscr{R}$ and the resulting map $\mathrm{Gal}(L/K) \to \mathfrak{S}(\mathscr{R})$ is an injective morphism of groups.

(We denote by $\mathfrak{S}(\mathscr{R})$ the group of permutations of the set $\mathfrak{R}$.)

Proof. If σ belongs to $\mathrm{Gal}(L/K)$ and $x \in L$, one has $\sigma(P(x)) = P(\sigma(x))$. In particular, if $P(x) = 0$, $P(\sigma(x)) = 0$, which means $\sigma(x) \in \mathscr{R}$. Since $\mathscr{R}$ is stable under $\sigma \in \mathrm{Gal}(L/K)$, the restriction of σ to $\mathscr{R}$ is an injective map $\mathscr{R} \to \mathscr{R}$. But $\mathscr{R}$ is finite, so $\sigma|_{\mathscr{R}}$ has to be bijective, hence is a permutation of $\mathscr{R}$. It is then clear that the map $\sigma \mapsto \sigma|_{\mathscr{R}}$ is a group morphism from $\mathrm{Gal}(L/K)$ to $\mathfrak{S}(\mathscr{R})$.

Let $\sigma \in \mathrm{Gal}(L/K)$ such that for any $x \in \mathscr{R}$, $\sigma(x) = x$. We need to prove that σ is the identity. The set L^σ consisting of all $a \in L$ such that $\sigma(a) = a$ is a subextension of L. Since it contains the roots of P and since, by assumption, L is generated by these roots, $L^\sigma = L$, which means $\sigma(a) = a$ for any $a \in L$. This shows that $\sigma = \mathrm{id}_L$, q.e.d. □

Recall that one says a group G acting on a set X acts *transitively* if for any x and $y \in X$, there is $g \in G$ such that $g \cdot x = y$.

Proposition 3.3.2. *Let K be a field, $P \in K[X]$ a separable polynomial and let $K \subset L$ be a splitting extension of P. The action of* $\mathrm{Gal}(L/K)$ *on the roots of P is transitive if and only if the polynomial P is irreducible in $K[X]$.*

Proof. Denote by $\mathscr{R}$ the set of roots of P in L. If P is not irreducible, there are two nonconstant polynomials Q and $R \in K[X]$ such that $P = QR$. Since P is separable, Q and R are coprime. It follows that the set $\mathscr{R}$ can be split into the union of the roots $\mathscr{R}_1$ of Q and $\mathscr{R}_2$ of R. The sets $\mathscr{R}_1$ and $\mathscr{R}_2$ are disjoint and nonempty. For $x \in \mathscr{R}_1$ and $\sigma \in \mathrm{Gal}(L/K)$, $Q(\sigma(x)) = \sigma(Q(x)) = \sigma(0) = 0$, so $\sigma(x) \in \mathscr{R}_1$. In particular, $\sigma(x) \notin \mathscr{R}_2$. For $x_1 \in \mathscr{R}_1$ and $x_2 \in \mathscr{R}_2$, this implies that there is no $\sigma \in \mathrm{Gal}(L/K)$ such that $\sigma(x_1) = x_2$.

Assume conversely that P is irreducible and let x, y be two roots of P. The subextensions $K \subset K[x]$ and $K \subset K[y]$ are both simple and generated by a root of P, so there is by Theorem 2.1.5 a unique K-morphism $f\colon K[x] \to K[y]$ such that $f(x) = y$. Since $K[x]$ and $K[y]$ have the same degree over K, f is an isomorphism. The field L is then a splitting extension of the polynomial P

over the two isomorphic subfields $K[x]$ and $K[y]$. By Theorem 2.2.2, one can extend f to a K-automorphism $\sigma\colon L \to L$. We just found an element σ in $\mathrm{Gal}(L/K)$ with $\sigma(x) = y$. □

As illustrated by the proof of Proposition 3.3.2, the explicit determination of elements in a Galois group is not very easy, since one has to work step by step. Each of these steps is, however, relatively manageable because it concerns only simple extensions. The next theorem states that, in fact, any finite separable extension is a simple extension.

Theorem 3.3.3 (Primitive Element Theorem). *Let K be a field and let $K \subset L$ be a finite separable extension. There exists $x \in L$ such that $L = K[x]$.*

Such an element x is usually called a *primitive element.*

Proof. One has to split the proof into two cases, according to the field K being finite or infinite.

Assume first that K is finite. Then the field L is finite, too, and its multiplicative group, L^*, is cyclic (Exercise 1.16). If x is a generator of L^*, one clearly has $L = K[x]$.

Assume now that K is infinite. By Corollary 3.2.11, there are only finitely many fields E with $K \subset E \subset L$. In particular, there are only finitely many simple extensions of K contained in L, say $E_1, \dots, E_n$.

Any element of L belongs to some E_i; indeed, any element x belongs to the simple extension $K[x]$, which, by assumption, is one of the E_i. This means that the K-vector space L is the union of finitely many vector subspaces. By the following lemma, there is some i with $E_i = L$, so that L is a simple extension of K. □

Lemma 3.3.4. *Let K be an infinite field, let V be a K-vector space and let $V_1, \dots, V_n$ be any finite family of vector subspaces in V such that $V_i \neq V$ for all i. Then $\bigcup_{i=1}^{n} V_i \neq V$.*

Proof. The result is obvious when $n = 1$. Let us prove it by induction on n. One may assume $V_1 \cup \dots \cup V_{n-1} \neq V$ and henceforth find $x \in V$ such that if $i \leqslant n-1$, $x \notin V_i$. If $x \notin V_n$, we are done. If not, since $V_n \neq V$, there is $y \in V$ with $y \notin V_n$. We will prove that there is $t \in K$ such that $x + ty$ does not belong to any V_i.

If $t \neq t'$ are such that both vectors $x+ty$ and $x+t'y$ belong to V_i, then the difference $(t-t')y \in V_i$, so $y \in V_i$ and consequently $x = (x+ty) - ty \in V_i$. If $i < n$, this contradicts the choice of x, and if $i = n$, this contradicts the choice of y. Therefore, for any i, there is at most one value of t for which $x + ty \in V_i$.

Hence there are at most n values of $t \in K$ for which $x + ty$ belongs to some V_i. Since K is infinite, one may choose $t \in K$ such that $x + ty \notin \bigcup_{i=1}^{n} V_i$. □

The method I chose for presenting Galois theory centers on the concept of morphism of extensions, and the crucial result was Artin's lemma (Proposition 3.2.8).

There is another approach which consists in first proving the Primitive Element Theorem (Theorem 3.3.3) without the use of Galois theory. The point is now that the Galois group of a simple extension is very easy to grasp: if $L = K[x]$, with x a root of an irreducible polynomial $P \in K[X]$ which is split in L, any K-automorphism of L is determined by the image of x and this image is one of the roots of P in L.

To make a connection between these two approaches to Galois theory, I will give you a direct proof of Theorem 3.3.3.

Another proof of the Primitive Element Theorem. The case of extensions of finite fields is treated as before, so we assume that the field K is infinite. By induction, it suffices to show that an extension which is generated by two elements is already generated by a single one. So let K be an infinite field, L a finite separable extension of K and x, y two elements in L such that $L = K[x, y]$. Let P and Q be the minimal polynomials of x and y over K. Let Ω be an algebraic closure of L and let us denote by $x = x_1, \dots, x_n$ (resp. $y = y_1, \dots, y_m$) the distinct roots of P (resp. Q) in Ω. Since the set of all $(x_i - x_1)/(y_j - y_1)$, for $1 \leqslant i \leqslant n$ and $1 \leqslant j \leqslant m$, is finite, and since K is infinite, we may find $c \in K$ such that for every $(i, j) \neq (1, 1)$, $x_i + cy_j \neq x_1 + cy_1$.

Set $z = x + cy$ and let us show that $L = K[z]$. The polynomial $R(X) = P(z - cX)$ has coefficients in the field $K[z]$ and vanishes at y. On the other hand, if $j \neq 1$, $z - cy_j = x_1 + cy_1 - cy_j$ is not a root of P, by the very choice of c, so $R(y_j) \neq 0$. It follows that y is the only root shared by Q and R. Since Q has simple roots, one has $\operatorname{g.c.d.}(Q, R) = X - y$. Now, Q and R both have coefficients in $K[z]$, hence so does their greatest common divisor (Proposition 2.4.3), and $y \in K[z]$. Finally $x = z - cy \in K[z]$, from which we conclude that $L = K[z]$. □

Thanks to this proof, most inductions become useless, at least if we stick to the study of separable extensions. Artin's lemma also may be given another proof.

Another proof of Proposition 3.2.8. It begins similarly. For $x \in L$, let us introduce its orbit under the action of G, that is, the set $\mathscr{O}_x$ of $\sigma(x)$ for $\sigma \in G$. The polynomial

$$P_x = \prod_{y \in \mathscr{O}_x} (X - y)$$

satisfies $\sigma(P_x) = P_x$, for any $\sigma \in G$, because any G-orbit is stable under σ. Therefore, the coefficients of P_x are elements of the field $K = L^G$. Since it has simple roots and since $P_x(x) = 0$, x is separable over K. It follows that the extension $K \subset L$ is separable. Moreover, the degree of any element of L over K is at most $\operatorname{card} G$.

Let us show that the extension $K \subset L$ is finite. Otherwise, there would exist an infinite increasing chain of finite subextensions, $K \subset L_1 \subset L_2 \subset \cdots \subset L$, with $[L_n : K] \to \infty$ when $n \to \infty$. By the Primitive Element Theorem, any extension $K \subset L_n$, being finite and separable, is a simple extension, hence $[L_n : K] \leqslant \operatorname{card} G$. This contradiction shows that $K \to L$ is a finite extension. By the Primitive Element Theorem again, it is a simple extension, hence $[L : K] \leqslant \operatorname{card} G$. Since by Proposition 3.2.4, one has $\operatorname{card} G \leqslant [L : K]$, the desired equality follows. □

3.4 Discriminant; resolvent polynomials

We shall begin this section by characterizing polynomials $P \in K[X]$ whose Galois group is contained in the alternating group. (By abuse of language, the Galois group of a separable polynomial is defined as the Galois group of some splitting extension.) This fits perfectly in the framework of Galois theory. Let L denote a splitting extension of a separable polynomial $P \in K[X]$. Choose an ordering $x_1, \ldots, x_n$ of the roots of P in L. The action of $\operatorname{Gal}(L/K)$ on $\{x_1, \ldots, x_n\}$ can now be transferred to an action of $\operatorname{Gal}(L/K)$ on $\{1, \ldots, n\}$, defined by

$$\sigma(x_i) = x_{\sigma(i)}, \quad 1 \leqslant i \leqslant n.$$

Now, any subgroup $H \subset \mathfrak{S}_n$ defines a subgroup $H \cap \operatorname{Gal}(L/K)$ of $\operatorname{Gal}(L/K)$ which corresponds to some subextension $K \subset L^{H \cap \operatorname{Gal}(L/K)} \subset L$. One can ask in particular whether $\operatorname{Gal}(L/K) \subset H$ or not.

But had we chosen another ordering $y_1, \ldots, y_n$ of the roots, there would exist some permutation $\tau \in \mathfrak{S}_n$ such that $y_i = x_{\tau(i)}$. The relations

$$\sigma(y_i) = \sigma(x_{\tau(i)}) = x_{\sigma\tau(i)} = y_{\tau^{-1}\sigma\tau(i)}$$

show that the subgroup G of $\mathfrak{S}_n$ defined by $\operatorname{Gal}(L/K)$ would be replaced by the conjugate subgroup $\tau^{-1} G \tau$. The preceding question must therefore be replaced by the following one: Is $\operatorname{Gal}(L/K)$ contained in a conjugate of H or not?

This problem does not appear for the alternating group $\mathfrak{A}_n \subset \mathfrak{S}_n$, since it is a normal subgroup. In other words, the property that $\operatorname{Gal}(L/K)$ is contained in $\mathfrak{A}_n$ does not depend on the ordering of the roots.

Recall that we had defined in Section 1.5 the discriminant as the symmetric polynomial in n variables

$$D(X_1, \dots, X_n) = \prod_{i<j} (X_i - X_j)^2.$$

By Theorem 1.5.3 on symmetric polynomials, there exists a unique polynomial $\Delta \in \mathbf{Z}[S_1, \dots, S_n]$ such that

$$D(X_1, \dots, X_n) = \Delta(S_1, \dots, S_n).$$

We now define the discriminant of a polynomial of degree n,

$$P = a_n X^n + a_{n-1} X^{n-1} + \dots + a_0, \quad a_n \neq 0,$$

by the formula

$$\operatorname{disc}(P) = a_n^{2n-2} \Delta(-a_{n-1}/a_n, a_{n-2}/a_n, \dots, (-1)^n a_0/a_n).$$

(The coefficient a_n^{2n-2} is there only for homogeneity, to insure, e.g. , that the discriminant of a polynomial with integer coefficients is again an integer; see Exercise 3.26.) If $x_1, \dots, x_n$ are roots of P in an algebraic closure of K, one thus has

$$\operatorname{disc}(P) = a_n^{2n-2} D(x_1, \dots, x_n).$$

In particular, $\operatorname{disc}(P) \neq 0$ if and only if P is separable.

Examples 3.4.1. a) The roots x and y of $P = aX^2 + bX + c$ satisfy $x + y = -b/a$ and $xy = c/a$, so that

$$\operatorname{disc}(P) = a^2(x-y)^2 = a^2(x^2 + y^2 - 2xy) = a^2((x+y)^2 - 4xy) = b^2 - 4ac.$$

b) With notations as above, one has

$$\operatorname{disc}(P) = (-1)^{n(n-1)/2} a_n^{n-2} \prod_{j=1}^{n} P'(x_j).$$

If we denote by $y_1, \dots, y_{n-1}$ the roots of P', it follows that

$$P'(x_j) = n a_n \prod_{k=1}^{n-1} (x_j - y_k),$$

and

$$\operatorname{disc}(P) = (-1)^{n(n-1)/2} a_n^{2n-2} n^n \prod_{k=1}^{n-1} \prod_{j=1}^{n} (x_j - y_k).$$

Therefore,

$$\operatorname{disc}(P) = (-1)^{n(n-1)/2} a_n^{n-1} n^n \prod_{k=1}^{n-1} P(y_k).$$

c) Look at the degree 3 polynomial defined by $P = X^3 + pX + q$, assuming that the characteristic of K is not equal to 3. One has $P' = 3X^2 + p$ and its roots are $\pm\sqrt{-p/3}$; hence

$$\begin{aligned}\operatorname{disc}(P) &= -27P(-\sqrt{-p/3})P(\sqrt{-p/3}) \\ &= -27\left(-\sqrt{-p/3}\frac{2p}{3}+q\right)\left(\sqrt{-p/3}\frac{2p}{3}+q\right) \\ &= -27q^2 - 4p^3.\end{aligned}$$

If P is a monic separable degree n polynomial, with roots $\{x_1, \ldots, x_n\}$ in some splitting extension L of K, set

$$d = \prod_{i<j}(x_i - x_j).$$

This is a square root in L of the discriminant of P. Moreover, for any $\sigma \in \operatorname{Gal}(L/K)$,

$$\sigma(d) = \prod_{i<j}(\sigma(x_i) - \sigma(x_j)) = \prod_{i<j}(x_{\sigma(i)} - x_{\sigma(j)}),$$

where $\sigma(i)$ is the unique integer in $\{1, \ldots, n\}$ such that $\sigma(x_i) = x_{\sigma(i)}$. Consequently,

$$\sigma(d)/d = \prod_{i<j}\begin{cases}1 & \text{if } \sigma(i) < \sigma(j); \\ -1 & \text{if } \sigma(i) > \sigma(j).\end{cases}$$

In other words, if $i(\sigma)$ denotes the number of inversions of the permutation σ acting on $\{1, \ldots, n\}$, then

$$\sigma(d)/d = (-1)^{i(\sigma)}$$

is equal to the *signature* of σ, viewed as an element of K. It follows that $d \in K$ if and only if the signature of any element in $\operatorname{Gal}(L/K)$ is equal to 1 in K, hence if and only if

- either the field K has characteristic 2 (in which case $-1 = 1$ in K!), or
- $\operatorname{Gal}(L/K)$ is a subgroup of the alternating group $\mathfrak{A}_n$.

This proves the following proposition, a concrete statement in Galois theory.

Proposition 3.4.2. *Let K be a field whose characteristic is different from 2. The Galois group of a monic separable polynomial P of degree n is contained in the alternating group A_n of the roots if and only if the discriminant of P is a square in K.*

Let P be a separable monic polynomial with coefficients in a field K, and let $K \subset L$ be a splitting extension of P. Let us denote by $x_1, \ldots, x_n$ the roots of P in L. For any polynomial f in $K[X_1, \ldots, X_n]$, let us consider the quantity $f(x_1, \ldots, x_n) \in L$. By Theorem 1.5.3, any symmetric polynomial with coefficients in K is a polynomial with coefficients in K in the elementary symmetric polynomials $S_1, \ldots, S_n$. As the $S_j(x_1, \ldots, x_n)$ are, up to sign, coefficients of the polynomial P, it follows that, for any symmetric polynomial $f \in K[X_1, \ldots, X_n]$, $f(x_1, \ldots, x_n)$ belongs to K.

More generally, assuming that $f(X_{\sigma(1)}, \ldots, X_{\sigma(n)}) = f(X_1, \ldots, X_n)$ for every permutation σ in a subgroup G of $\mathfrak{S}_n$, it will follow that $f(x_1, \ldots, x_n) \in K$, as soon as $\operatorname{Gal}(L/K) \subset G$. This motivates the following considerations.

Let G be any subgroup of $\mathfrak{S}_n$. Let f be a polynomial in $\mathbf{Z}[X_1, \ldots, X_n]$ and let H be the subgroup of G consisting of all $\sigma \in G$ such that

$$f(X_{\sigma(1)}, \ldots, X_{\sigma(n)}) = f(X_1, \ldots, X_n).$$

In fact, the formula ${}^{\sigma}f = f(X_{\sigma(1)}, \ldots, X_{\sigma(n)})$ defines an action $[\sigma]\colon f \mapsto {}^{\sigma}f$ of the symmetric group $\mathfrak{S}_n$ on the polynomial ring $K[X_1, \ldots, X_n]$. Observe that any $[\sigma]$ is both an automorphism of rings and of K-vector spaces. If we restrict this action to the subgroup G, then the group H is precisely the stabilizer of the polynomial f. Therefore, for $\sigma \in G$ and $\tau \in H$, ${}^{\sigma\tau}f = {}^{\sigma}({}^{\tau}f) = {}^{\sigma}f$ and ${}^{\sigma}f$ only depends on the right coset σH of σ in G/H. As a consequence, $f(x_{\sigma(1)}, \ldots, x_{\sigma(n)})$ still depends only on this coset σH, which allows the following definition to make sense.

Definition 3.4.3. *The* resolvent polynomial $R_G(f, P)$ *of* P *with respect to* f *and* G *is defined by*

$$R_G(f, P) = \prod_{\sigma H \in G/H} (X - f(x_{\sigma(1)}, \ldots, x_{\sigma(n)})).$$

(The product is on an arbitrary family of elements in G representing cosets in G/H.)

Lemma 3.4.4. *If* $\operatorname{Gal}(L/K)$ *is a subgroup of* G, $R_G(f, P) \in K[X]$.

Proof. By definition, $R_G(f, P)$ is a polynomial in $L[X]$. To prove that it belongs to $K[X]$, one has to check that for any element $\tau \in \operatorname{Gal}(L/K)$, $\tau(R_G(f, P)) = R_G(f, P)$. (The action of $\operatorname{Gal}(L/K)$ on polynomials is defined coefficient-wise; a polynomial is stable under $\operatorname{Gal}(L/K)$ if and only if each of its coefficients is.) But, by the very definition of the map $\operatorname{Gal}(L/K) \to \mathfrak{S}_n$, one has $\tau(x_j) = x_{\tau(j)}$ for every j, so that

$$\tau(f(x_{\sigma(1)}, \ldots, x_{\sigma(n)})) = f(x_{\tau\sigma(1)}, \ldots, x_{\tau\sigma(n)})$$

and it follows that

$$\begin{aligned}\tau(R_G(f,P)) &= \tau\Big(\prod_{\sigma\in G/H}(X - f(x_{\sigma(1)},\dots,x_{\sigma(n)}))\Big)\\ &= \prod_{\sigma\in G/H}(X - f(x_{\tau\sigma(1)},\dots,x_{\tau\sigma(n)})).\end{aligned}$$

Since we assumed that $\mathrm{Gal}(L/K)\subset G$, when $[\sigma]$ runs along the set of right cosets in G modulo H so does $[\tau\sigma]$, hence $\tau(R_G(f,P)) = R_G(f,P)$, q.e.d. □

The discriminant appears to be a particular case of this general setup of resolvent polynomials. In fact, the stabilizer of the polynomial $f = \prod_{i<j}(X_i - X_j)$ in $G = \mathfrak{S}_n$ is equal to the alternating group $H = \mathfrak{A}_n$. One has $f(x_1,\dots,x_n) = d$, while if σ is any odd permutation, $f(x_{\sigma(1)},\dots,x_{\sigma(n)}) = -d$. Consequently, $R_{\mathfrak{S}_n}(f,P) = (X-d)(X+d) = X^2 - d^2 = X^2 - \mathrm{disc}(P)$ and the criterion of Prop. 3.4.2 can be rewritten as follows: the Galois group of P is contained in $\mathfrak{A}_n$ if and only if the resolvent polynomial $R_{\mathfrak{S}_n}(f,P)$ has a simple root in K. (Observe that the polynomial $X^2 - a$ is split separable if and only if a is a nonzero square and the characteristic of the field is not equal to 2.)

For a more general resolvent polynomial, one can prove the following proposition.

Proposition 3.4.5. *Keeping the previous notation, assume that* $\mathrm{Gal}(L/K)\subset G$*, and assume moreover that the resolvent polynomial* $R_G(f,P)\in K[X]$ *has a simple root in* K*. Then,* $\mathrm{Gal}(L/K)$ *is* conjugate *in* G *to a subgroup of* H*. In other words, there is* $g\in G$ *such that* $\mathrm{Gal}(L/K)\subset gHg^{-1}$.

Proof. For any $\sigma\in G$, set $\alpha_\sigma = f(x_{\sigma(1)},\dots,x_{\sigma(n)})$. It was established in the proof of Lemma 3.4.4 that for any $\tau\in\mathrm{Gal}(L/K)$, one has $\tau(\alpha_\sigma) = \alpha_{\tau\sigma}$. Let $\sigma\in G$ be any element such that the simple root whose existence is claimed by the statement is α_σ (the right coset σH of the element σ is well defined, although σ is not). That α_σ belongs to K means that for any $\tau\in\mathrm{Gal}(L/K)$, one has $\alpha_{\tau\sigma} = \tau(\alpha_\sigma) = \alpha_\sigma$. Since α_σ is a simple root of $R_G(f,P)$, if follows that for any $\tau\in\mathrm{Gal}(L/K)$, $\tau\sigma$ is in the same right coset modulo H as σ. This means that $\tau\sigma\in\sigma H$, that is, $\tau\in\sigma H\sigma^{-1}$, which is exactly what we had to prove. □

A generalization of this proposition, together with an adequate list of (G,f) is actually used by computer programs to determine explicitly Galois groups of polynomials with integer coefficients, at least when their degree is not too big. (Present knowledge and computer capacities go to degree 23, in MAGMA's implementation.)

3.5 Finite fields

A finite field is a field with finitely many elements. An interesting aspect of finite fields is that they illustrate in a rather clear and simple manner Galois theory, in addition to being an incredibly rich topic in arithmetic. They were discovered by Gauss but he did not publish anything on this subject, and Galois rediscovered them later.

Let us begin with some easy remarks. Let F be a finite field, and let q denote its cardinality. The canonical morphism $\mathbf{Z} \to F$ cannot be injective, so that the characteristic of F is a prime number p. In particular, F contains the field $\mathbf{F}_p = \mathbf{Z}/p\mathbf{Z}$, and is naturally a $\mathbf{F}_p$-vector space, whose dimension is necessarily finite. If d denotes this dimension, it follows that $\operatorname{card} F = p^d$.

The set F^* of nonzero elements in F is a group under multiplication, with cardinality $q-1$. By Lagrange's theorem (Proposition 4.2.2), any element $x \in F^*$ satisfies $x^{q-1} = 1$. Equivalently, $x^q = x$ for any $x \in F$. Like any finite subgroup of the multiplicative group of a field, F^* is cyclic (see Exercise 1.16) and there exists an element $x_1 \in F^*$ whose order is $q-1$.

If $F \to \Omega$ is an embedding of F in an algebraic closure of $\mathbf{F}_p$, the elements of F map to roots of the equation $x^q = x$. Conversely, the set of elements $x \in \Omega$ satisfying $x^q = x$ is a finite set, and its cardinality is exactly q, because the polynomial $X^q - X$, having derivative -1, is separable. Moreover, F is a subfield of Ω. If φ is the Frobenius homomorphism of Ω, given by $\varphi(x) = x^p$, F is the subfield fixed by the automorphism $\varphi^d \colon x \mapsto x^{p^d}$.

This description also shows that two finite fields with the same cardinality q are isomorphic: they are both isomorphic to the subfield of Ω consisting of all x with $x^q = x$.

In brief:

Proposition 3.5.1. *Let p be a prime number, and let $q = p^d$ be a power of p. The roots of the polynomial $X^q - X$ in an algebraic closure of $\mathbf{F}_p$ form a finite field with q elements. It is isomorphic to any field with q elements.*

Let us now look at extensions of finite fields. Let $E \to F$ be such an extension. Necessarily, the fields E and F have the same characteristic p and there are integers e and f such that $\operatorname{card} E = p^e$ and $\operatorname{card} F = p^f$. Since F is a finite-dimensional E-vector space, the cardinality of F is a power of that of E. This shows that e divides f. Moreover $d = f/e$ is the degree of the extension $E \to F$. Conversely, let E and F be two finite fields with cardinalities p^e and p^f, where $f = de$. If Ω is an algebraic closure of $\mathbf{F}_p$, E and F can be identified with the subfields of Ω consisting of elements x with $x^{p^e} = x$ and $x^{p^f} = x$, respectively. This allows us to identify E with a subfield of F, hence there is some morphism $E \to F$.

Without loss of generality, we now assume that $E \subset F$. Let $\Phi = \varphi^e : F \to F$ denote the morphism of fields given by $x \mapsto x^{p^e}$. It restricts to the identity on E, and so defines an element of $\mathrm{Aut}(F/E)$. If $x_1 \in F^*$ has order $p^f - 1$, one has $\Phi^j(x) = x^{p^{ej}}$, hence $\Phi^j(x) \neq x$ for $1 \leqslant ej < f$. This shows that the d elements $\mathrm{id}, \Phi, \dots, \Phi^{d-1}$ in $\mathrm{Aut}(F/E)$ are distinct, hence $\operatorname{card} \mathrm{Aut}(F/E) \geqslant d$. Consequently, the extension $E \to F$ is Galois, and its Galois group is generated by Φ, whence is isomorphic to $\mathbf{Z}/d\mathbf{Z}$.

We have just proved:

Proposition 3.5.2. *Extensions of finite fields are Galois, and their corresponding Galois groups are cyclic.*

This fact has an important and concrete application to the action of the Galois group on the roots of a polynomial with coefficients in a finite field. To simplify, let us consider a separable monic polynomial $P \in \mathbf{F}_p[X]$ and let F be a splitting extension of P. Let $\varphi\colon x \mapsto x^p$ be the Frobenius homomorphism, which is a generator of the Galois group $\mathrm{Gal}(F/\mathbf{F}_p)$. If x is a root of P, so is $\varphi(x)$. We can order the roots $\{x_1, x_2, \dots, x_n\}$ of P so that the permutation that φ induces on them has the decomposition into cycles with disjoint supports

$$(x_1, \dots, x_{n_1})(x_{n_1+1}, \dots, x_{n_1+n_2}) \dots (x_{n_1+\dots+n_{r-1}+1}, \dots, x_n).$$

(There are a r cycles, with lengths $n_1, \dots, n_r$, and $n_1 + \dots + n_r = n$.) Since $\mathrm{Gal}(F/\mathbf{F}_p)$ is generated by φ, the roots of P in a given cycle are exactly the conjugates of any of them, hence a factorization of the polynomial P into irreducible factors, with degrees $n_1, \dots, n_r$.

On the other hand, the degrees of the irreducible factors of the polynomial P allow one to determine the length of the cycles of the Frobenius morphism acting on the roots of P.

As last remark, to compute the n_j, one does not really need to factor the polynomial P, for it is sufficient to determine the number ν_d of irreducible factors with given degree d. By definition, ν_1 is the number of roots (distinct since P is separable) of P in $\mathbf{F}_p$; these are the common roots of P and $X^p - X$, so that ν_1 is the degree of the polynomial $\mathrm{g.c.d.}(P, X^p - X)$. Then, if $\mathbf{F}_{p^2}$ denotes an extension of $\mathbf{F}_p$ with cardinality p^2, $\nu_1 + 2\nu_2$ is equal to the number of roots of P in the field $\mathbf{F}_{p^2}$. This implies that $2\nu_2$ is the degree of the g.c.d. of the polynomials P and $(X^{p^2} - X)/(X^p - X)$, which is also equal to the g.c.d. of $P/\mathrm{g.c.d.}(P, X^p - X)$ and $X^{p^2} - X$. More generally, the number of roots of P in the field $\mathbf{F}_{p^n}$ is equal to

$$\sum_{d \mid n} d\nu_d = \deg \gcd(X^{p^n} - X, P).$$

This allows us to compute ν_d by induction.

Exercises

Exercise 3.1 (Extending algebraic identities). Let K be an infinite field.

a) If $P \in K[X_1, \dots, X_n]$ is a nonzero polynomial, show that there is $(x_1, \dots, x_n) \in K^n$ with $P(x_1, \dots, x_n) \neq 0$.

b) Prove that the polynomial ring $K[X_1, \dots, X_n]$ is an integral domain. Deduce from this another proof for Lemma 3.3.4.

c) Let P and Q be two polynomials in $K[X_1, \dots, X_n]$. Assume that $Q \neq 0$ and that for any $(x_1, \dots, x_n) \in K^n$ with $Q(x_1, \dots, x_n) \neq 0$, one has $P(x_1, \dots, x_n) = 0$. Show that $P = 0$.

Exercise 3.2. Let K be a field.

a) Show that a polynomial $P \in K[X]$ is separable if and only if it is a product of distinct *irreducible* separable polynomials.

b) Let $P_1, \dots, P_r$ be separable polynomials in $K[X]$. Show that their l.c.m. and their g.c.d. are separable polynomials.

Exercise 3.3. Let K be the field generated by i and $\sqrt{2}$ in $\mathbf{C}$ ($i^2 = -1$).

a) Show that $[K : \mathbf{Q}] = 4$. Write down a primitive element for K and compute its minimal polynomial.

b) Determine all possible actions of $\mathrm{Gal}(K/\mathbf{Q})$ on the set $\{\pm i, \pm\sqrt{2}\}$. Deduce that $\mathrm{Gal}(K/\mathbf{Q}) \simeq (\mathbf{Z}/2\mathbf{Z})^2$.

c) Determine all subfields of K.

Exercise 3.4. Let k be a field and let K denote the field $k(X)$ of rational functions with coefficients in k.

a) Let $R \in K$ be any nonconstant rational function, and let L denote the subfield $k(R)$ of K generated by R over k. Write $R = A/B$ as the quotient of two coprime polynomials. Show that the polynomial $P(X, Y) = A(X) - B(X)Y$ is irreducible in $k[X, Y]$.

b) (*continued*) Conclude that X is algebraic over L, with degree $\max(\deg A, \deg B)$.

c) Let f be any k-automorphism of $k(X)$. Show that there is a matrix $\left(\begin{smallmatrix} a & b \\ c & d \end{smallmatrix}\right)$ in $\mathrm{GL}(2, k)$ such that $f(X) = (aX + b)/(cX + d)$.

d) Show that $\mathrm{Aut}(k(X)/k)$ is isomorphic the group $\mathrm{PGL}(2, k) = \mathrm{GL}(2, k)/k^*$.

Exercise 3.5. Let $K \subset L$ be a Galois extension with Galois group G. Let H be a subgroup of G and let $E = L^H$ be the extension of K fixed by H. Show that the Galois closure E^{g} of E is contained in L and determine the subgroup of G to which it corresponds by Galois theory.

Exercise 3.6. Let K be a field and let L be a finite Galois extension of K. Let $x \in L$ and let y be a conjugate of x, that is, a root of the minimal polynomial of x over K.

Show that there exists $\sigma \in \mathrm{Gal}(L/K)$ with $\sigma(x) = y$. How many such σ are there?

Exercise 3.7. Let K be a field and let $K \subset L$ be a Galois extension which is a splitting extension of some irreducible polynomial $P \in K[X]$. Let $n = \deg P$ and denote by $x_1, \dots, x_n$ its roots in L. This allows us to consider $\mathrm{Gal}(L/K)$ as a subgroup of $\mathfrak{S}_n$.

a) For $i \in \{1, \dots, n\}$, let H_i be the subgroup of $\mathfrak{S}_n$ of permutations σ with $\sigma(i) = i$. Set $G_i = \mathrm{Gal}(L/K) \cap H_i$. Show that the G_i are conjugate in $\mathrm{Gal}(L/K)$ and that for any i, $(\mathrm{Gal}(L/K) : G_i) = n$. To which subextensions do these subgroups correspond?

b) If $\mathrm{Gal}(L/K)$ is abelian, show that $L = K[x_1]$.

Exercise 3.8. Let $K \subset L$ be a finite Galois extension. Let $x \in L$ and assume that the elements $\sigma(x)$ for $\sigma \in \mathrm{Gal}(L/K)$ are all distinct. Show that $L = K(x)$.

Exercise 3.9. Let $K \to L$ be a simple extension, and fix $\alpha \in L$ such that $L = K[\alpha]$. For any field E, with $K \subset E \subset L$, let P_E denote the minimal polynomial of α over E.

a) Show that $[L : E] = \deg P_E$.

b) If E, E' are two fields such that $K \subset E' \subset E \subset L$, show that P_E divides $P_{E'}$.

c) Let E be a field with $K \subset E \subset L$, and let E' be the subfield of E generated by the coefficients of P_E. Show that $P_{E'} = P_E$. Conclude that $E' = E$.

d) Show that there are only finitely many fields E with $K \subset E \subset L$.

Exercise 3.10. **a)** Let $K \to L$ be an algebraic separable extension. Let n be an integer such that any element in L has degree at most n over K. Show that $[L : K] \leqslant n$. (*Otherwise, there would be a finite extension $L_1 \subset L$ with $[L_1 : K] > n$. Apply the Primitive Element Theorem.*)

b) Let p be a prime number. Set $K = \mathbf{F}_p(X, Y)$ and let L be the extension of K generated by the pth roots of X and Y in an algebraic closure of K. Show that $[L : K] = p^2$ but that any element in L has degree at most p over K. In particular, the extension $K \subset L$ cannot be simple.

Exercise 3.11. Let k be a field, $L = k(X_1, \dots, X_n)$ the field of rational functions in n variables and let K be the subfield of L generated by k and the elementary symmetric polynomials $S_1, \dots, S_n$.

a) Show that the extension $K \subset L$ is Galois with group $\mathfrak{S}_n$.

b) If k has characteristic zero, show that $X_1 + 2X_2 + \cdots + nX_n$ generates L over K.

c) Let $f \in L$, denote by H its stabilizer in $\mathfrak{S}_n$, i.e. , the set of all $\sigma \in \mathfrak{S}_n$ such that

$$f(X_1, \dots, X_n) = f(X_{\sigma(1)}, \dots, X_{\sigma(n)}).$$

Show that the extension $K(f) \subset L$ is Galois with group H. (Prove that f is a primitive element of the extension $K \subset L^H$.)

d) Deduce from the previous question that any rational function $g \in L$ with the same stabilizer as f can be expressed as a rational function in f and the elementary symmetric polynomials (Lagrange, 1770, 60 years before Galois!).

Explicit example with $n = 3$: $f = X_1 X_2 + X_3$, $g = X_3$.

Exercise 3.12. **a)** Let G be a group and let F be a field. Let $\sigma_1, \ldots, \sigma_n$ be n distinct group morphisms from G to the multiplicative group $F^\times$. Show that $\sigma_1, \ldots, \sigma_n$ are linearly independent: if $a_1, \ldots, a_n$ are elements in F with $a_1\sigma_1 + \cdots + a_n\sigma_n = 0$, then $a_1 = \cdots = a_n = 0$. (*Argue by contradiction, considering a nontrivial relation with the least possible number of nonzero coefficients.*)

b) Let E and F be two fields and let $\sigma_1, \ldots, \sigma_n$ be n distinct field morphisms $E \to F$. Deduce from the first question that they are linearly independent over F.

Exercise 3.13. Let $K \to \Omega$ be an algebraic extension of a perfect field K and assume that any nonconstant polynomial in $K[X]$ has a root in Ω.

a) Let $K \to L$ be a splitting extension of an irreducible polynomial $P \in K[X]$. Show that there is $\alpha \in L$ such that $L = K(\alpha)$.

b) Let Q denote the minimal polynomial of α over K. Using the fact that Q has a root in Ω, show that there is a morphism of extensions $L \to \Omega$.

c) Conclude that P is split in Ω, hence that Ω is an algebraic closure of K.

Exercise 3.14. Let $K = \mathbf{Q}(\zeta)$ be the extension of $\mathbf{Q}$ generated by a primitive nth root of unity ζ in $\mathbf{C}$.

a) Show that the set of nth roots of unity in K is a cyclic group of order n, generated by ζ.

b) Show that the extension $\mathbf{Q} \to K$ is Galois.

c) Let σ be an element of $\mathrm{Gal}(K/\mathbf{Q})$. Show that there exists an integer d prime to n such that $\sigma(\zeta) = \zeta^d$.

d) Construct an injective group morphism $\varphi\colon \mathrm{Gal}(K/\mathbf{Q}) \to (\mathbf{Z}/n\mathbf{Z})^*$.

e) Using Exercise 2.5, prove that φ is an isomorphism of groups.

Exercise 3.15. Let p be an odd prime number. Recall that $(\mathbf{Z}/p\mathbf{Z})^*$ is a cyclic group with $p-1$ elements.

a) Show that for any integer $n \geqslant 1$, $(\mathbf{Z}/p^n\mathbf{Z})^*$ is a cyclic group of order $(p-1)p^{n-1}$. (If $a \in \mathbf{Z}$ is an element of the class modulo p which generates $(\mathbf{Z}/p\mathbf{Z})^*$, consider $a^{p^n}(1+p)$.)

b) Conclude that for any positive integer n, the cyclotomic field $\mathbf{Q}(\zeta_{p^n})$ generated by a primitive p^nth root of unity has a unique subfield K with $[K : \mathbf{Q}] = 2$.

c) Justify the existence of a group morphism $\varepsilon\colon (\mathbf{Z}/p\mathbf{Z})^* \to \{\pm 1\}$ such that any generator of $(\mathbf{Z}/p\mathbf{Z})^*$ maps to -1. One then defines

$$G = \sum_{k=1}^{p-1} \varepsilon(k) \exp(2ik\pi/p)$$

(*Gauss's sum*). Show that $G^2 = (-1)^{(p-1)/2}p$. Can you now identify the subfield asserted by the previous question?

Exercise 3.16. Let $F = \mathbf{F}_q$ be a finite field with q elements. If $d \geqslant 1$, let $\mathscr{I}_d$ denote the set of irreducible monic irreducible polynomials of degree d in $F[X]$; let $N_d = \operatorname{card} \mathscr{I}_d$.

a) Show that for any positive integer d, $N_d \neq 0$.

b) For any positive integer n, show that n is a multiple of the degree of any irreducible polynomial dividing $X^{q^n} - X$. Conversely, if d divides n, show that any polynomial in $\mathscr{I}_d$ divides $X^{q^n} - X$.

c) Show that

$$q^n = \sum_{d|n} dN_d.$$

d) Show that for any positive integer n,

$$N_n \geqslant \frac{1}{n} q^n \frac{q-2}{q-1}.$$

e) Let $\mu\colon \mathbf{N}^* \to \{0, 1, -1\}$ denote Möbius function, defined by $\mu(n) = (-1)^r$ if n is the product of r distinct prime numbers, and $\mu(n) = 0$ otherwise. For any two functions f, $g\colon \mathbf{N}^* \to \mathbf{C}$, show that $f(n) = \sum_{d|n} g(d)$ for all n, if and only if $g(n) = \sum_{d|n} \mu(n/d) f(d)$ for all n (*Möbius's inversion formula*). In particular,

$$N_n = \frac{1}{n} \sum_{d|n} \mu(n/d) q^d.$$

Exercise 3.17. This exercise focuses on the factorization of cyclotomic polynomials over a finite field $\mathbf{F}_q$ with cardinality q and characteristic p. Let Ω be an algebraic closure of $\mathbf{F}_q$.

a) For any positive integer n which is prime to p, and any nonnegative integer r, show that $\Phi_{p^r n} = \Phi_n^{p^r - 1}$ in the ring $\mathbf{F}_q[X]$.

b) Let $\alpha \in \Omega^*$. Show that there exists a least positive integer n such that $\alpha^n = 1$. Show that it is prime to p. Show that the degree of α over $\mathbf{F}_q$ is equal to the order of q in the group $(\mathbf{Z}/n\mathbf{Z})^*$.

c) Let $n \geqslant 2$ be an integer prime to p. Show that the polynomial Φ_n is separable in $\mathbf{F}_q[X]$. Deduce from the preceding question that all of these irreducible factors in $\mathbf{F}_q[X]$ have the same degree.

d) Show that the polynomial $X^4 + 1$ is irreducible in $\mathbf{Q}[X]$ but that it is reducible in $\mathbf{F}_p[X]$ for any prime number p. Generalize.

Exercise 3.18. a) Let n and a be two integers, and let p be a prime number dividing $\Phi_n(a)$ (Φ_n denotes the nth cyclotomic polynomial). Show that $p \equiv 1 \pmod{n}$ unless p divides n.

b) Let n be any positive integer. Show that there are infinitely many prime numbers of the form $kn + 1$, for $k \in \mathbf{N}$. (*This is an elementary special case of Dirichlet's theorem according to which, if n and m are two coprime integers, there are infinitely many prime numbers of the form $kn + m$.*)

Exercise 3.19. Let $\mathbf{F}$ be a finite field, q its cardinality and p its characteristic. For $f \in \mathbf{F}[x_1, \dots, x_n]$, set $S(f) = \sum_{x \in \mathbf{F}^n} f(x)$.

a) Compute $S(x_1^{i_1} \dots x_n^{i_n})$ in terms of $(i_1, \dots, i_n)$.

b) Let $f_1, \dots, f_r$ be polynomials in $\mathbf{F}[x_1, \dots, x_n]$ with degrees $d_1, \dots, d_r$, and let V be the set of their common zeroes in $\mathbf{F}^n$.

Let f be the polynomial defined by $f = \prod_{i=1}^{r} (1 - f_i^{q-1})$. Compute $S(f)$ in terms of $\operatorname{card} V$. Deduce that if $d_1 + \dots + d_r < n$, $\operatorname{card} V$ is divisible by p (*Chevalley–Warning's theorem*).

Exercise 3.20. Let $\mathbf{F}$ be a finite field, and denote its cardinality by q.

a) Show that the vector space $\mathbf{F}^2$ is the union of $q+1$ lines, and that $q+1$ lines are indeed necessary to cover $\mathbf{F}^2$. More generally, how many hyperplanes are needed to cover $\mathbf{F}^n$?

b) Let $H_1, \dots, H_d$ be affine hyperplanes in $\mathbf{F}^n$ which do not pass through the origin and which cover $\mathbf{F}^n \setminus \{0\}$. Show that $d \geqslant n(q-1)$. Show also that this lower bound is optimal by exhibiting such a cover with $d = n(q-1)$. (*If f_i is an equation for H_i, set $f = \prod_{i=1}^{d} f_i$ and consider the quantity $S(f)$ defined in Exercise 3.19.*)

Exercise 3.21. This exercise is the basis of Berlekamp's algorithm for factoring polynomials over finite fields.

Let P be a nonconstant separable polynomial with coefficients in the finite field $\mathbf{F}_p$. We denote by R_P the ring $\mathbf{F}_p[X]/(P)$. Let $P = \prod_{i=1}^{r} P_i$ be the factorization of P in irreducible polynomials in $\mathbf{F}_p[X]$. Let $n_i = \deg P_i$.

a) Show that the ring R_{P_i} is isomorphic to the finite field $\mathbf{F}_{p^{n_i}}$.

b) If $A \in R_P$, let $\rho_i(A)$ be the remainder in the Euclidean division of A by P_i. Show that the map $A \mapsto (\rho_1(A), \dots, \rho_r(A))$ is an isomorphism of rings $R_P \simeq \prod_{i=1}^{r} R_{P_i}$.

c) For $A \in R_P$, let $t(A) = A^p - A$. Show that t is a $\mathbf{F}_p$-linear endomorphism of R_P (viewed as a $\mathbf{F}_p$-vector space) which corresponds, by the previous isomorphisms, to the map

$$\prod_{i=1}^{r} \mathbf{F}_{p^{n_i}} \to \prod_{i=1}^{r} \mathbf{F}_{p^{n_i}}, \qquad (a_1, \dots, a_r) \mapsto (a_1^p - a_1, \dots, a_r^p - a_r).$$

d) Show that the kernel of t is a $\mathbf{F}_p$-vector subspace of R_P of dimension r.

e) Let a be an element in the kernel of t. Show that there is a monic polynomial $Q \in \mathbf{F}_p[X]$ of minimal degree such that $Q(a) = 0$. Show that the polynomial Q is separable and split over $\mathbf{F}_p$.

f) (*continued*) If $a \notin \mathbf{F}_p$, show that Q is not irreducible. From a partial factorization $Q = Q_1 Q_2$, show how to deduce a nontrivial partial factorization of P.

Exercise 3.22. **a)** For any integer $n \geqslant 1$, compute the discriminant of the polynomial $X^n - 1$.

b) Let p and q be two elements in a field K. Show that the polynomial $X^5 + pX + q$ has discriminant equal to $5^5 q^4 + 4^4 p^5$.

c) Generalize the previous questions by computing, for any positive integer n, the discriminant of the polynomial $X^n + pX + q$.

Exercise 3.23. Let p and q be two distinct prime numbers. Let F be a splitting extension of the polynomial $P = X^q - 1$ over the field $\mathbf{F}_p$.

a) The Frobenius morphism $\varphi \in \operatorname{Gal}(F/\mathbf{F}_p)$ induces a permutation of the roots of P. Determine its decomposition into cycles with disjoint supports.

b) Show that $\operatorname{Gal}(F/\mathbf{F}_p)$ is a subgroup of the alternating group $\mathfrak{A}_q$ if and only if p is a square in $\mathbf{Z}/q\mathbf{Z}$.

Exercise 3.24 (Quadratic reciprocity). If $a \in (\mathbf{Z}/q\mathbf{Z})^*$, one sets $\left(\frac{a}{q}\right) = 1$ if a is a square in $\mathbf{Z}/q\mathbf{Z}$, otherwise $\left(\frac{a}{q}\right) = -1$ (*Legendre's symbol*).

a) If q is odd, show that $\left(\frac{a}{q}\right) = a^{(q-1)/2}$.

b) Using results from Exercise 3.22 and 3.23, prove the *quadratic reciprocity law* (Gauss, April 8, 1796); if p and q are two distinct odd prime numbers,

$$\left(\frac{p}{q}\right)\left(\frac{q}{p}\right) = (-1)^{(p-1)(q-1)/4}.$$

c) Let p be an odd prime number. Choose a primitive 8th root of unity ζ in an algebraic closure of $\mathbf{F}_p$. Set $\alpha = \zeta + \zeta^{-1}$.

Compute α^2. Deduce that 2 is a square in $\mathbf{F}_p$ if and only if $p \equiv \pm 1 \pmod 8$. Check finally that

$$\left(\frac{2}{p}\right) = (-1)^{(p^2-1)/8}.$$

Exercise 3.25 (Resultants and discriminant). Let

$$A = a_n(X - x_1)\dots(X - x_n) = a_n X^n + \dots + a_0,$$
$$B = b_m(X - y_1)\dots(X - y_m) = b_m X^m + \dots + b_0$$

be two split polynomials with coefficients in a field K.

a) Consider the following matrices with coefficients in K:

$$S = \begin{pmatrix} a_n & a_{n-1} & \dots & a_0 & & & \\ & \ddots & & & \ddots & & \\ & & & a_n & a_{n-1} & \dots & a_0 \\ b_m & b_{m-1} & \dots & b_0 & & & \\ & \ddots & & & \ddots & & \\ & & & b_m & b_{m-1} & \dots & b_0 \end{pmatrix},$$

where there are m lines of a's and then n lines of b's, and

$$V = \begin{pmatrix} y_1^{n+m-1} & \dots & y_m^{n+m-1} & x_1^{n+m-1} & \dots & x_n^{n+m-1} \\ \vdots & & \vdots & \vdots & & \vdots \\ y_1 & \dots & y_m & x_1 & \dots & x_n \\ 1 & \dots & 1 & 1 & \dots & 1 \end{pmatrix}.$$

The determinant of S is called the *resultant* of A and B and denoted $\operatorname{Res}(A, B)$. By computing the product SV and taking determinants, show that

$$\operatorname{Res}(A, B) = a_n^m b_m^n \prod_{\substack{1 \leqslant i \leqslant n \\ 1 \leqslant j \leqslant m}} (x_i - y_j) = a_n^m \prod_{i=1}^{n} B(x_i) = (-1)^{mn} b_m^n \prod_{j=1}^{m} A(y_j),$$

at least when the polynomial AB has only simple roots.

b) Using Exercise 3.1, show that the previous formulae always hold.

c) If $m = n - 1$ and $B = A'$, show that

$$\operatorname{Res}(A, A') = (-1)^{n(n-1)/2} a_n \operatorname{disc}(A).$$

Exercise 3.26. Let $P = a_n X^n + a_{n-1} X^{n-1} + \cdots + a_0$ be a polynomial of degree n with coefficients in $\mathbf{Z}$. Show that $\operatorname{disc}(P)$ is an integer. (Use either Exercise 1.18 or Exercise 3.25).

4

A bit of group theory

This chapter explains essential notions in group theory that one uses in Galois theory. This chapter was really taught according to the needs of students and should be read likewise.

4.1 Groups (quick review of basic definitions)

A *group* is a set G endowed with an internal law $(g, g') \mapsto g * g'$ (also called the product of g and g') satisfying the following properties:

– there is an element $e \in G$ such that for any $g \in G$, $e * g = g * e = g$ (*identity element*);

– for any $g \in G$, there is $g' \in G$ such that $g * g' = g' * g = e$ (any element has an *inverse*);

– for any g, g', g'' in G, one has $g * (g' * g'') = (g * g') * g''$ (*associativity*).

Numerous notations exist for the internal law besides $*$, for example, $\cdot$, $\times$, $+$, $\bullet$, $*$, ., etc. If there is no risk of confusion, it is common to omit the symbol and to simply use gg' to denote the *product* of two elements g and g' in a group G. One often denotes the inverse of an element g by g^{-1}, especially when the product is denoted $\cdot$. Similarly, the identity element might be denoted e_G (to distinguish between groups), or 1, or 1_G.

The additive notation $+$ is used only for *commutative* groups, that is, groups such that for any elements g and g', $gg' = g'g$. In that case, the identity element is denoted 0, or 0_G, and the inverse of an element g is denoted $-g$. These groups are also called abelian, in honor of the Norwegian mathematician Niels Henryk Abel.

Here are some *examples of groups*: the group $\mathfrak{S}_n$ of permutations of the set $\{1, \ldots, n\}$ (the product being composition), the group $\mathbf{Z}$ of integers (with addition), the set of nonzero real numbers (with multiplication), any vector space (with addition), the set of $n\times n$ invertible matrices (with multiplication), and the set of orthogonal $n \times n$ matrices (again with multiplication).

4.2 Subgroups

If G and H are two groups, a *group homomorphism* $f\colon G \to H$ is a map f such that $f(gg') = f(g)f(g')$ for any g and g' in G. If $f : G \to H$ is a homomorphism, $f(e_G) = e_H$ and for any $g \in G$, $f(g^{-1}) = f(g)^{-1}$. An *isomorphism* is a bijective homomorphism.

Let G be a group. A set $H \subset G$ is a *subgroup* of G if $e_G \in H$, the product of any two elements in H is in H, and if the inverse of any element of H is in H. Then the product law of G restricts to an internal law on H, giving H the structure of a group with identity e_G. If S is a subset of a group G, $\langle S\rangle$ denotes the subgroup of G *generated* by S, that is, the smallest subgroup of G containing S.

Lemma 4.2.1. *Let G be a group and let S be a nonempty subset of G. The subgroup $\langle S\rangle$ is the set of all products $s_1 \dots s_n$, where $s_i \in S$ or $s_i^{-1} \in S$ for any i.*

Proof. Let H be the set of all such products. Since any subgroup of G containing S must contains these products, one has $H \subset \langle S\rangle$. Conversely, $S \subset H$, hence it it is sufficient to show that H is a group. But the inverse of $s_1 \dots s_n \in H$ is equal to $s_n^{-1} \dots s_1^{-1}$, so belongs to H. Similarly, if $s_1 \dots s_n \in H$ and $t_1 \dots t_m \in H$, then their product $s_1 \dots s_n t_1 \dots t_m \in H$. Moreover, $e_G \in H$ (given either by the product of an empty family of elements, or by writing it ss^{-1} for any element $s \in S$). □

The image $f(G)$ of a morphism of groups $f\colon G \to H$ is a subgroup in H. The preimage $f^{-1}(H')$ of any subgroup $H' \subset H$ is a subgroup of G. In particular, the *kernel* of f is the subgroup of all $g \in G$ with $f(g) = e_H$. It is equal to $\{e_G\}$ if and only if f is injective.

If a group G is finite, we use the word *order* as a synonym for its cardinality. The order of an element $g \in G$ is the smallest integer $n \geqslant 1$ such that $g^n = e$ (if there is no such integer, one says that g has infinite order). This is also the order of the subgroup $\langle g\rangle$ generated by g in G. Observe that the group $\langle g\rangle$ generated by g is isomorphic to $\mathbf{Z}/n\mathbf{Z}$ if g has finite order n, and is isomorphic to $\mathbf{Z}$ if g has infinite order (*exercise*).

Proposition 4.2.2 (Lagrange). *Let G be a finite group and let H be a subgroup of G. The order of H divides that of G. In particular, the order of any element in G divides the order of the group G.*

The *index* of a subgroup H of a finite group G is the quotient $\operatorname{card} G / \operatorname{card} H$ and is denoted $(G : H)$.

Proof. Let us define a relation in G by setting $g \sim g'$ if there exists $h \in H$ such that $g' = gh$. This is an *equivalence relation:*

– reflexivity: since $g = ge$ and $e \in H$, $g \sim g$ for any $g \in G$;

– symmetry: if $g \sim g'$, let $h \in H$ be such that $g' = gh$. Then, $g = g'h^{-1}$, hence $g' \sim g$ since $h^{-1} \in H$;

– transitivity: if $g \sim g'$ and $g' \sim g''$, let h and $h' \in H$ be such that $g' = gh$ and $g'' = g'h'$. One then has $g'' = g(hh')$. Since H is a subgroup, $hh' \in H$ and $g \sim g''$.

The *equivalence class* of an element $g \in G$ is the set gH of all gh for $h \in H$. Since G is a group, the map $h \mapsto gh$ defines a bijection from H to gH: all equivalence classes are in bijection. In particular, if G is finite, they all have the same cardinality, that of H.

The group G is the disjoint union of its equivalence classes for this relation $\sim$. If there are N classes, then $\operatorname{card} G = N \operatorname{card} H$ and the order of H divides the order of G. □

The classes gH defined above are called *right cosets*[1] of H in G. One denotes by G/H the set of all right cosets. Similarly, one defines another equivalence relation: $g \sim g'$ if $g' = hg$. Equivalence classes are now sets Hg for $g \in G$ and the set of all these *left cosets* is denoted by $H \backslash G$. When G is finite, the two sets G/H and $H \backslash G$ have the same cardinality: the index $(G : H)$ of H in G.

There is a partial converse to Lagrange's theorem, attributed to Cauchy.

Proposition 4.2.3 (Cauchy's lemma). *Let G be a finite group and let p be a prime number that divides the order of G. Then there is an element in G whose order is equal to p.*

Proof. When G is abelian, this is an immediate consequence of Exercise 1.15. We then prove the proposition by induction on the order of G.

Let Z be the *center* of G, that is, the set of all $g \in G$ such that $gh = hg$ for any $h \in G$. This is a commutative subgroup of G. If p divides the order

[1] Right cosets are orbits for the action of H by translation on the right in G but some authors like Lang [7] or Bourbaki call them left cosets.

of Z, one is done, either by induction if $\operatorname{card} Z < \operatorname{card} G$, or by the case of an abelian group if $\operatorname{card} Z = \operatorname{card} G$.

We say that two elements x and y of G are *conjugate* if there is $g \in G$ such that $y = gxg^{-1}$. This is an equivalence relation in G. For $x \in G$, let us denote by $\mathscr{C}_x$ its *conjugacy class*, that is, the set of all $y \in G$ that are conjugate to x. The map $G \to \mathscr{C}_x$ defined by $g \mapsto gxg^{-1}$ is surjective. Moreover, if $gxg^{-1} = hxh^{-1}$, then $y = h^{-1}g$ satisfies $yx = xy$, and conversely. The set of all such y defines a subgroup G_x in G, called the *centralizer* of x in G. Hence, the cardinality of $\mathscr{C}_x$ multiplied by that of G_x is equal to the order of G:

$$\operatorname{card} \mathscr{C}_x = (G : G_x).$$

Now the group G is the union of disjoint conjugacy classes: there are elements $x_1, \dots, x_r \in G$ such that no two of them are conjugate, and such that any element in G is conjugate to one of them. One has therefore the following so-called *class formula*:

$$\operatorname{card} G = \sum_{i=1}^{r} (G : G_{x_i}).$$

Note that an element in Z is conjugate only to itself. That means that all elements in Z appear in this sum, so that we may write

$$\operatorname{card} G = \operatorname{card} Z + \sum_{\substack{i=1 \\ x_i \notin Z}}^{r} (G : G_{x_i}).$$

Since $\operatorname{card} G$ is divisible by p while $\operatorname{card} Z$ is not, there must be some i with $x_i \notin Z$ such that $(G : G_{x_i})$ is prime to p. This implies that p divides $\operatorname{card} G_{x_i}$. The condition $x_i \notin Z$ means that $G_{x_i} \neq G$, hence $\operatorname{card} G_{x_i} < \operatorname{card} G$, and we conclude by induction. □

4.3 Group actions

An *action* of a group G on a set X is a group morphism of G to the group $\mathfrak{S}(X)$ of permutations of X. In numerous cases, there is no possible confusion as to the morphism in question and one denotes by $g \cdot x$, or even gx, the image of $x \in X$ by the permutation of $\mathfrak{S}(X)$ associated to g.

The *orbit* $\mathscr{O}_x$ of an element $x \in X$ is the set of all gx for g in G. The *stabilizer* $\operatorname{Stab}_G(x)$ of x is the set of all $h \in G$ with $hx = x$. This is a subgroup of G. Moreover one has $gx = g'x$ if and only if $g'g^{-1}x = x$, that is, $g^{-1}g' \in \operatorname{Stab}_G(x)$, or again $g' \in g \operatorname{Stab}_G(x)$. In other words, the map $G \to X$

given by $g \mapsto gx$ induces a bijection from the set $G/\operatorname{Stab}_G(x)$ of right cosets of $\operatorname{Stab}_G(x)$ to the orbit $\mathscr{O}_x$ of x.

In particular, if a finite group G acts on a set X, one has for any $x \in X$ the equality

$$\operatorname{card} \mathscr{O}_x = (G : \operatorname{Stab}_G(x)) = \frac{\operatorname{card} G}{\operatorname{card} \operatorname{Stab}_G(x)}.$$

The set X is the union of all its orbits. Moreover, for any x and $y \in X$, either their orbits are disjoint, or they are equal. If we now pick one element x_i per orbit, the set X is the disjoint union of the orbits $\mathscr{O}_{x_i}$. If X is also finite, its cardinality is the sum of the cardinalities of each $\mathscr{O}_{x_i}$, hence

$$\boxed{\operatorname{card} X = \sum_i \operatorname{card} \mathscr{O}_{x_i} = \sum_i \frac{\operatorname{card} G}{\operatorname{card} \operatorname{Stab}_G(x_i)}.}$$

This is the *class formula*. It is very important since, applied to a well-chosen action, it imposes quite a strong condition on the orders of various subgroups.

In fact, the proof of Proposition 4.2.3 already made use of this class formula, applied to the *conjugation action*, which is the action of the group G on itself defined by $g \cdot h = ghg^{-1}$. The centralizer of an element $g \in G$ is nothing but the stabilizer of g for this action.

Left and right translations in a group are two other important examples of actions of a group G on itself, defined by $g \cdot h = gh$ and $g \cdot h = hg^{-1}$, respectively. An action of a group restricts naturally to an action of any of its subgroups. Therefore a subgroup $H \subset G$ acts on G by left or right translations; it is easy to check that orbits of these actions are just left and right cosets.

4.4 Normal subgroups; quotient groups

The notion of *normal subgroup* that we now introduce appears naturally when one wants to define *quotients* in the category of groups. Not every subgroup can be the kernel of a group morphism. Indeed, let $\varphi\colon G \to G'$ be a group homomorphism; for any g, $h \in G$, one has

$$\varphi(g^{-1}hg) = \varphi(g)^{-1}\varphi(h)\varphi(g).$$

In particular, if $\varphi(h) = e$, one has $\varphi(g^{-1}hg) = \varphi(g)^{-1}e\varphi(g) = e$. Hence the definition:

Definition 4.4.1. *A subgroup H in a group G is said to be* normal *if for any $g \in G$ and any $h \in H$, $g^{-1}hg \in H$.*

Proposition 4.4.2. *The kernel of any group homomorphism is a normal subgroup.*

Examples 4.4.3. *a*) Let G be a group. The trivial subgroups $\{1\}$ and G are normal subgroups of G.

b) Subgroups of a commutative group are automatically normal.

c) Let Z be the center of a group G (the set of $g \in G$ such that for any $h \in G$, $gh = hg$). This is a normal subgroup of G, for if $h \in Z$ and $g \in G$, $g^{-1}hg = g^{-1}gh = h \in Z$.

d) Let G be a group and let D be the subgroup of G (*derived subgroup*) generated by *commutators*, i.e. , expressions of the form $g_1g_2g_1^{-1}g_2^{-1}$ with g_1 and g_2 in G. This is a normal subgroup of G. Take $g \in G$ and consider a commutator $c = g_1g_2g_1^{-1}g_2^{-1}$. One has

$$\begin{aligned} g^{-1}cg &= g^{-1}(g_1g_2g_1^{-1}g_2^{-1})g \\ &= (g^{-1}g_1g)(g^{-1}g_2g)(g^{-1}g_1^{-1}g)(g^{-1}g_2^{-1}g) \\ &= (g^{-1}g_1g)(g^{-1}g_2g)(g^{-1}g_1g)^{-1}(g^{-1}g_2g)^{-1} \\ &= h_1h_2h_1^{-1}h_2^{-1}, \end{aligned}$$

with $h_1 = g^{-1}g_1g$ and $h_2 = g^{-1}h_1g$, so $g^{-1}cg$ is still a commutator. Since the inverse of a commutator is again a commutator, any element d in D can be written as $d = d_1 \dots d_n$ for some commutators d_i. The formula

$$g^{-1}dg = (g^{-1}d_1g)\dots(g^{-1}d_ng)$$

and the above remark show that $g^{-1}dg$ is a product of commutators, so belong to D, q.e.d.

Exercise 4.4.4. Show that a subgroup $H \subset G$ is a normal subgroup if and only if $gH = Hg$ for any $g \in G$.

The construction of a *quotient group* of a group by a normal subgroup will show that any normal subgroup is the kernel of a canonical group morphism. Let us consider the set G/H of right cosets of H and let us try to define a group law in G/H: for g and $g' \in G$, h and $h \in H$, one has

$$(gh)(g'h') = gg'(g')^{-1}hg'h'$$

and, since H is a normal subgroup of G, $(g')^{-1}hg'$ belongs to H, hence so does $(g')^{-1}hg'h'$. This shows that the right coset $(gg')H$ depends only on the right cosets gH and $g'H$, but not on the actual elements g and g'. We thus may set

$$(gH) * (g'H) = (gg')H.$$

Now that we have explained why this definition makes sense, it is a routine exercise to check that it endows G/H with a group structure whose identity is the coset $H = eH$, and that the map $G \to G/H$ given by $g \mapsto gH$ is a surjective group morphism. By construction, its kernel is H.

One virtue of quotient groups is the following factorization theorem (a universal property, again).

Theorem 4.4.5. *Let G be a group, let H be a normal subgroup of G, and denote by $\pi\colon G \to G/H$ the canonical morphism $g \mapsto gH$. Let $f\colon G \to G'$ be a group morphism whose kernel contains H. Then there exists a unique group morphism $\varphi\colon G/H \to G'$ such that for any $g \in G$, $\varphi(\pi(g)) = f(g)$. (In other words, $\varphi \circ \pi = f$.)*

The kernel of φ is equal to $\pi(\operatorname{Ker} f)$. In particular, φ is injective if and only if $\operatorname{Ker} f = H$. Finally, φ is surjective if and only if f is.

Proof. If $\pi(g) = \pi(g')$, there exists $h \in H$ such that $g = g'h$, which implies $f(g) = f(g'h) = f(g')f(h) = f(g')$ since $H \subset \operatorname{Ker} f$. In particular, we can define a map $G/H \to G'$ by associating to a coset in G/H the image by f of any element in this coset, because this image will not depend on the chosen element. Let us call this map φ; by construction, the identity $\varphi \circ \pi = f$ holds.

It is now routine to check that $\varphi\colon G/H \to G'$ is a group morphism. A coset gH belongs to $\operatorname{Ker}\varphi$ if and only if $f(g) = e$, that is, if and only if $g \in \operatorname{Ker} f$. This proves that $\operatorname{Ker}\varphi = \pi(\operatorname{Ker} f)$.

Assume φ is injective and let $g \in \operatorname{Ker} f$; one has $\varphi(gH) = f(g)H = H$, hence $gH \in \operatorname{Ker}\varphi$ and $gH = H$. This implies $g \in H$, so that $\operatorname{Ker} f \subset H$, hence the equality. Conversely, if $\operatorname{Ker} f = H$, let $g \in G$ be such that $gH \in \operatorname{Ker}\varphi$. It follows that $g \in \operatorname{Ker} f$, hence $g \in H$ and $gH = H$ is the identity element of G/H, hence φ is injective. Finally, since the map $\pi\colon G \to G/H$ is surjective, surjectivity of f and of φ are clearly equivalent. □

Proposition 4.4.6. *Let G be a group, let H be a normal subgroup of G and let $\pi\colon G \to G/H$ denote the canonical morphism with kernel H.*

a) *For any subgroup K in G/H, $\pi^{-1}(K)$ is a subgroup of G that contains H, and all such subgroups are obtained in this way.*

b) *If K is a normal subgroup of G/H, then the composition $G \to G/H \to (G/H)/K$ is a group morphism with kernel $\pi^{-1}(K)$. It induces an isomorphism*

$$G/\pi^{-1}(K) \simeq (G/H)/K.$$

Proof. a) Like any preimage, $\pi^{-1}(K)$ is a subgroup of G. Moreover, it contains $\pi^{-1}(e) = H$. Conversely, let K be a subgroup of G containing H. Then $\pi(K)$ is a subgroup of G/H and $\pi^{-1}(\pi(K))$ contains K. To show the other inclusion, let us consider $g \in \pi^{-1}(\pi(K))$. It follows that $\pi(g) \in \pi(K)$, so that there is $k \in K$ with $\pi(g) = \pi(k)$. Consequently, $h = gk^{-1}$ is an element of $\operatorname{Ker}\pi = H$. Since $H \subset K$, $g = hk$ belongs to K.

b) The given map $f\colon G \to (G/H)/K$ is a surjective morphism of groups, being the composition of two surjective group morphisms. Moreover, an element $g \in G$ belongs to $\operatorname{Ker} f$ if and only if $\pi(g)$ belongs to the kernel K of the morphism $G/H \to (G/H)/K$, i.e. $\operatorname{Ker} f = \pi^{-1}(K)$. It follows that f induces an isomorphism $G/\pi^{-1}(K) \simeq (G/H)/K$. □

Definition 4.4.7. *One says that a group G is* simple *if its only normal subgroups are $\{e\}$ and G.*

4.5 Solvable groups; nilpotent groups

Let G be a group. If G has a normal subgroup N, one can in a sense "approximate" the group structure of G by the combination of those of N and G/N. More generally, it may be worth introducing a *normal series* in G, that is a sequence of subgroups

$$\{1\} = G_0 \subset G_1 \subset \cdots \subset G_{n-1} \subset G_n = G$$

of G such that for any integer i, $1 \leqslant i \leqslant n$, G_{i-1} is a normal subgroup of G_i.

Definition 4.5.1. *A group G is* solvable *if it has a normal series*

$$\{1\} = G_0 \subset G_1 \subset \cdots \subset G_{n-1} \subset G_n = G$$

such that for any i, $1 \leqslant i \leqslant n$, the quotient group G_i/G_{i-1} a commutative group.

This notion fits very well with subgroups and quotients:

Proposition 4.5.2. *Let G be a group, and let H be a subgroup of G.*

a) *If G is solvable, then H is solvable.*
b) *If G is solvable and H is normal, then G/H is solvable.*
c) *If H is solvable and normal, and if G/H is solvable, then G is solvable.*

Proof. a) By assumption, there are subgroups $G_0 \subset \ldots G_n$ with G_{i-1} normal in G_i and G_i/G_{i-1} abelian. Set $H_i = H \cap G_i$. The subgroups H_i form an increasing sequence of subgroups in H. Let us consider the restriction to H_i of the canonical map $G_i \to G_i/G_{i-1}$ which is a group morphism. Its kernel is equal to $G_{i-1} \cap H_i = H_{i-1}$. Therefore H_{i-1} is a normal subgroup of H_i and the induced map $H_i/H_{i-1} \to G_i/G_{i-1}$ given by Theorem 4.4.5 is injective, so identifies the group H_i/H_{i-1} with a sugroup of G_i/G_{-1}. In particular, H_i/H_{i-1} is abelian. We thus have shown that H is solvable.

b) Assume moreover that H is a normal subgroup of G and let us show that G/H is solvable. Let $\pi\colon G \to G/H$ be the canonical morphism and for any i, let us set $K_i = \pi(G_i)$. Since π is onto, it is easy to check that the group K_{i-1} is normal in K_i; the induced morphism $G_i \to K_i \to K_i/K_{i-1}$ is surjective and its kernel contains G_{i-1}. By Theorem 4.4.5, we deduce the existence of a surjective morphism $G_i/G_{i-1} \to K_i/K_{i-1}$. Since G_i/G_{i-1} is abelian, K_i/K_{i-1} is abelian too. This shows that G/H is solvable.

c) Since H is solvable, we may find subgroups

$$\{1\} = H_0 \subset H_1 \subset \cdots \subset H_m = H$$

with successive quotients H_i/H_{i-1} abelian groups. Since G/H is assumed to be solvable, there is an analogous series of subgroups in G/H:

$$\{1\} = K_0 \subset K_1 \subset \cdots \subset K_n = G/H.$$

In the series of subgroups in G,

$$\{1\} = H_0 \subset H_1 \subset \cdots \subset H_m = H = \pi^{-1}(K_0) \\ \subset \pi^{-1}(K_1) \subset \cdots \subset \pi^{-1}(K_n) = G,$$

any subgroup is a normal subgroup of the next, with quotient an abelian group. This is clear for the first m subgroups; for the last n subgroups, notice that, K_{i-1} being normal in K_i, the morphism

$$\pi^{-1}(K_i) \xrightarrow{\pi} K_i \to K_i/K_{i-1}$$

is surjective and its kernel equals $\pi^{-1}(K_{i-1})$. This shows that $\pi^{-1}(K_{i-1})$ is normal in $\pi^{-1}(K_i)$ and

$$\pi^{-1}(K_i)/\pi^{-1}(K_{i-1}) \simeq K_i/K_{i-1}$$

is abelian, as we needed to prove. □

For finite groups, to be solvable is equivalent to an apparently more restrictive notion:

Proposition 4.5.3. *Let G be a finite group. Then G is solvable if and only if G has a normal series*

$$\{e_G\} = G_0 \subset G_1 \subset \cdots \subset G_{n-1} \subset G_n = G$$

such that for any $i \in \{0, \dots, n-1\}$, G_{i+1}/G_i is cyclic.

In particular, if G is a finite solvable group, there exists a normal subgroup H in G such that G/H is isomorphic to $\mathbf{Z}/d\mathbf{Z}$, for some integer $d \geqslant 2$.

Proof. Since cyclic groups are commutative, a group satisfying this criterion is solvable.

For the converse, let us first show that a finite abelian group has such a normal series. If G is a finite abelian group, we can find elements $x_1, \dots, x_r \in G$ such that $G = \langle x_1, \dots, x_r \rangle$ (take for example all elements in G). Since G is commutative, any subgroup of G is normal. Consequently, the chain of subgroups

$$\{e_G\} \subset \langle x_1 \rangle \subset \langle x_1, x_2 \rangle \subset \cdots \subset \langle x_1, \dots, x_r \rangle = G$$

is a normal series in G. Moreover, the quotient $\langle x_1, \dots, x_j \rangle / \langle x_1, \dots, x_{j-1} \rangle$ is generated by x_j, so is cyclic. This shows that the proposition holds for abelian groups.

Let us now prove that any finite solvable group G has a normal series such as stated in the proposition. By induction, we may assume that this holds for finite solvable groups of cardinality $< \operatorname{card} G$. By hypothesis, G has a normal subgroup $H \neq G$ such that G/H is commutative. By induction, there is a normal series with cyclic quotients in H,

$$\{e_G\} \subset G_1 \subset \cdots \subset G_m = H.$$

By the abelian case, there is also a normal series with cyclic quotients in G/H, $\{e_{G/H}\} = K_0 \subset \cdots \subset K_r = G/H$. For $1 \leqslant i \leqslant r$, let G_{m+i} be the preimage of K_i in G. Then G_{m+i-1} is a normal subgroup of G_{m+i} and the quotient G_{m+i}/G_{m+i-1} is isomorphic to H_i/H_{i-1}, hence is cyclic. Finally, the chain of subgroups

$$\{e_G\} \subset G_1 \subset \cdots \subset G_m \subset G_{m+1} \subset \cdots \subset G_{m+r} = G$$

is a normal series with cyclic quotients, as required. □

A similar definition gives rise to a different, but interesting, notion.

Definition 4.5.4. *One says that a group G is* nilpotent *if it has a normal series*

$$\{1\} = G_0 \subset G_1 \subset \cdots \subset G_{n-1} \subset G_n = G$$

such that for any $i \in \{0, \dots, n-1\}$, G_i is a normal subgroup of G and G_{i+1}/G_i is contained in the center of G/G_i.

A nilpotent group is automatically solvable.

4.6 Symmetric and alternating groups

Recall that we denote by $\mathfrak{S}_n$ the group of all permutations of the set $\{1, 2, \dots, n\}$. Its cardinality is $n!$. A *transposition* in $\mathfrak{S}_n$ is a permutation that switches two distinct elements and fixes all the others. The *cycle* $(i_1, \dots, i_m)$ is the permutation such that $i_1 \mapsto i_2, \dots, i_{m-1} \mapsto i_m$ and $i_m \mapsto i_1$, all other elements being fixed. Its *length* is m. A transposition is thus a cycle with length 2.

In the symmetric group, it appears to be quite easy to decide if any two elements in $\mathfrak{S}_n$ are conjugate. Any element $\sigma \in \mathfrak{S}_n$ can be written as a product of cycles with disjoint supports, whose lengths form a *partition* $\pi(\sigma)$ of the integer n, that is a decomposition $n = i_1 + \dots + i_r$ with positive integers $i_1, \dots, i_r$. It is common usage to write them in increasing order.

Proposition 4.6.1. *Two permutations are conjugate if and only if they define the same partition.*

In particular, any two transpositions are conjugate in $\mathfrak{S}_n$, as are any two 3-cycles.

Proof. Let $(m_1, \dots, m_p)$ be the partition associated to σ and to τ. We can write

$$\sigma = (i_{1,1}, \dots, i_{1,m_1}) \dots (i_{p,1}, \dots, i_{p,m_p}),$$

for integers $i_{k,r}$. Likewise, we can find such integers for τ, say $j_{k,r}$. Let γ be the permutation mapping $i_{k,r}$ to $j_{k,r}$. Then, for $1 \leqslant r < m_k$, one has

$$\gamma\sigma\gamma^{-1}(j_{k,r}) = \gamma\sigma(i_{k,r}) = \gamma(i_{k,r+1}) = j_{k,r+1},$$

while

$$\gamma\sigma\gamma^{-1}(j_{k,m_k}) = \gamma\sigma(i_{k,m_k}) = \gamma(i_{k,1}) = j_{k,1}.$$

This shows that $\gamma\sigma\gamma^{-1} = \tau$, so that σ and τ are conjugate.

The same computation shows the converse. If σ is as above and γ is any element of $\mathfrak{S}_n$,

$$\gamma\sigma\gamma^{-1} = (\gamma(i_{1,1}), \dots, \gamma(i_{1,m_1})) \dots (\gamma(i_{p,1}), \dots, \gamma(i_{p,m_p})),$$

which is a product of cycles with disjoint supports and lengths $m_1, \dots, m_p$. Two conjugate elements thus define the same partition of n. □

To give more properties of the symmetric group, we will have to use the fact that it is generated by some very special subsets.

Proposition 4.6.2. *The group $\mathfrak{S}_n$ is generated by (at will):*

a) *the transpositions (i, j), for $1 \leqslant i, j \leqslant n$;*
b) *the transpositions $(i, i+1)$, for $1 \leqslant i \leqslant n-1$;*
c) *the transposition $(1, 2)$ and the cycle $(1, 2, \dots, n)$.*

Proof. a) We have to show that any permutation $\sigma \in \mathfrak{S}_n$ is a product of transpositions. Let us show this by induction on the least integer k such that σ fixes k, $k+1$, ..., n. If $k = 1$, σ is the identity, so the result is true. Then assume that σ fixes $k+1, \dots, n$, set $j = \sigma(k)$ and define $\tau = (j,k) \circ \sigma$. This is a permutation in $\mathfrak{S}_n$ fixing $k+1, \dots, n$ (note that $j \leqslant k$), and one has $\tau(k) = (j,k)(\sigma(k)) = (j,k)(j) = k$. By induction, τ is a product of transpositions, and $\sigma = (j,k) \circ \tau$ is a product of transpositions too.

b) Let H be the subgroup generated by the transpositions $(i, i+1)$, for $1 \leqslant i \leqslant n-1$. For any two elements $p < m$ in $\{1, \dots, n\}$, the product

$$\tau = (m-1, m) \circ (m-2, m-1) \circ (p, p+1)$$

belongs to H and satisfies $\tau(p) = m$, while τ fixes $m+1, \dots, n$. The same argument as for a) then shows that $H = \mathfrak{S}_n$.

c) Set $\tau = (1,2)$ and let γ be the cycle $(1, 2, \dots, n)$. We have to show that the subgroup H generated by τ and γ is equal to $\mathfrak{S}_n$. By a computation we did in the proof of Proposition 4.6.1,

$$\gamma\tau\gamma^{-1} = \gamma(1,2)\gamma^{-1} = (\gamma(1), \gamma(2)) = (2,3),$$

and similarly $\gamma^{i-1}\tau\gamma^{1-i} = (i, i+1)$ for any integer i, with $1 \leqslant i \leqslant n-1$. All these elements belong to the subgroup H. By *b)*, $H = \mathfrak{S}_n$. □

We shall derive from these facts a characterization of morphisms from $\mathfrak{S}_n$ to any commutative group. It will turn out that the *signature* is essentially the only such morphism.

To be complete, let us recall its definition and show that it is a group morphism from $\mathfrak{S}_n$ to $\{\pm 1\}$. The easiest way to define the signature is to set

$$\varepsilon(\sigma) = \prod_{1 \leqslant i < j \leqslant n} \frac{\sigma(i) - \sigma(j)}{i - j}.$$

This is a rational number with absolute value 1 (all unordered pairs $\{i,j\}$ appear both in the numerator and in the denominator), so is ± 1. More precisely, $\varepsilon(\sigma) = (-1)^{i(\sigma)}$, where $i(\sigma)$ denotes the number of inversions of the permutation σ, that is the number of pairs (i,j) with $i < j$ such that $\sigma(i) > \sigma(j)$.

One may check directly that $\varepsilon(\sigma\tau) = \varepsilon(\sigma)\varepsilon(\tau)$, but we nearly did so in Section 3.4 (page 70). Consider the polynomial $d = \prod_{i<j} (X_i - X_j)$. Then, denoting by ${}^\sigma f = f(X_{\sigma(1)}, \dots, X_{\sigma(n)})$ the action of a permutation $\sigma \in \mathfrak{S}_n$ on a polynomial $f \in \mathbf{Q}[X_1, \dots, X_n]$, we proved the equality ${}^\sigma d = \varepsilon(\sigma) d$. The fact that this $f \mapsto {}^\sigma f$ is an action implies that ε is a morphism of groups $\mathfrak{S}_n \to \{\pm 1\}$.

A permutation σ is said to be *even* or *odd*, according to whether $\varepsilon(\sigma) = 1$ or -1. Even permutations form a subgroup $\mathfrak{A}_n \subset \mathfrak{S}_n$, called the *alternating* group. This is the kernel of ε, hence a normal subgroup. The permutation $\sigma = (1, 2)$ has only one inversion, for the pair $(1, 2)$, hence $\varepsilon(\sigma) = -1$. In particular, the signature is a surjective morphism and the alternating group has index 2 in $\mathfrak{S}_n$. Observe also that the signature of any transposition is equal to -1, because any two transpositions are conjugate.

Proposition 4.6.3. *Let A be a commutative group (with composition law denoted additively) and let $f \colon \mathfrak{S}_n \to A$ be any group morphism. There exists a unique element $a \in A$ such that $2a = 0$ and such that for any $\sigma \in \mathfrak{S}_n$, $f(\sigma) = 0$ if $\varepsilon(\sigma) = 1$ and $f(\sigma) = a$ if $\varepsilon(\sigma) = -1$.*

Proof. Since two transpositions are conjuguate in $\mathfrak{S}_n$, they have the same image by f, say $a \in A$. A transposition τ satisfying $\tau^2 = \mathrm{id}$, one has $2a = 2f(\tau) = f(\tau^2) = f(\mathrm{id}) = 0$. On the other hand, if an element σ in $\mathfrak{S}_n$ is the product of m transpositions $\tau_1, \ldots, \tau_m$, then

$$f(\sigma) = f(\tau_1) + \cdots + f(\tau_m) = ma.$$

Observe that $\varepsilon(\sigma) = (-1)^m$. Consequently, if $\varepsilon(\sigma) = 1$, then m is even and $f(\sigma) = 0$, while if $\varepsilon(\sigma) = -1$, then m is odd and $f(\sigma) = a$. □

Corollary 4.6.4. *The derived subgroup of $\mathfrak{S}_n$ is the alternating group $\mathfrak{A}_n$.*

Proof. Any commutator is clearly an even permutation, therefore the derived subgroup $D(\mathfrak{S}_n)$ is contained in $\mathfrak{A}_n$. Conversely, since $D(\mathfrak{S}_n)$ is a normal subgroup of $\mathfrak{S}_n$, we may consider the quotient group $A = \mathfrak{S}_n/D(\mathfrak{S}_n)$. Any commutator in A is the image of a commutator in $\mathfrak{S}_n$ by the canonical surjective group morphism $\mathfrak{S}_n \to A$. Consequently, any commutator in A is trivial and the group A is abelian. By Proposition 4.6.3, the map $\mathfrak{S}_n \to A$ factors through the signature, hence its kernel $D(\mathfrak{S}_n)$ contains $\mathfrak{A}_n$. We therefore have equality. □

We shall play again this game in the alternating group. All will result from the three following properties.

- the 3-cycles generate $\mathfrak{A}_n$;
- the square of the 3-cycle $(1, 2, 3)$ is equal to the 3-cycle $(1, 3, 2)$;
- if $n \geqslant 5$, any two 3-cycles are conjugate in $\mathfrak{A}_n$.

Lemma 4.6.5. *The group $\mathfrak{A}_n$ is generated by the cycles of length* 3.

Proof. It suffices to show that the product of any two transpositions is a product of 3-cycles, for any element in $\mathfrak{A}_n$ is the product of an *even* number of transpositions. The formula $(1,2)(1,3) = (1,3,2)$ takes care of any product with one element in common, while the formula

$$(1,2)(3,4) = (1,2)(2,3)\,(2,3)(3,4) = (1,2,3)\,(2,4,3)$$

treats the case of two transpositions with disjoint supports. □

Proposition 4.6.6. *If $n \geqslant 5$, there is no nonzero morphism from the alternating group to an abelian group.*

Proof. Let us prove, as was claimed, that if $n \geqslant 5$, the cycles $\gamma = (1,2,3)$ and $\delta = (a,b,c)$ are conjugate in $\mathfrak{A}_n$. We may find a permutation σ in $\mathfrak{S}_n$ with $\sigma(1) = a$, $\sigma(2) = b$ and $\sigma(3) = c$. By an easy computation we did during the proof of Proposition 4.6.1,

$$\sigma\gamma\sigma^{-1} = \delta.$$

If σ belongs to $\mathfrak{A}_n$, then the two 3-cycles are conjugate in $\mathfrak{A}_n$, as we wanted to show. But if σ is odd, we can modify it. The idea is simple: it suffices to multiply σ with a transposition τ such that $\tau\gamma\tau^{-1} = \gamma$. Take for instance $\tau = (4,5)$. (Here we use the fact that $n \geqslant 5$: we have to consider two elements outside of the 3-cycle $(1,2,3)$.) Then one has

$$(\sigma\tau)\gamma(\sigma\tau)^{-1} = \sigma(\tau\gamma\tau^{-1})\sigma^{-1} = \sigma\gamma\sigma^{-1} = \delta,$$

which shows that γ and δ are conjugate in $\mathfrak{A}_n$.

Assume now that $n \geqslant 5$ and let $f\colon \mathfrak{A}_n \to A$ be a morphism to an abelian group A. Since all 3-cycles are conjugate, they have the same image in A. On the other hand, the square of the 3-cycle $(1,2,3)$ is equal to the 3-cycle $(1,3,2)$. This shows that $f((1,3,2)) = 2f((1,2,3))$, whence $a = 2a$ and $a = 0$. Therefore, for any 3-cycle $c \in \mathfrak{A}_n$, $f(c) = 0$. Since they generate $\mathfrak{A}_n$, $f = 0$. □

The next two corollaries follow immediately.

Corollary 4.6.7. *If $n \geqslant 5$, the derived subgroup of $\mathfrak{A}_n$ is $\mathfrak{A}_n$ itself.*

Corollary 4.6.8. *If $n \geqslant 5$, $\mathfrak{A}_n$ and $\mathfrak{S}_n$ are not solvable.*

Remark 4.6.9. Actually, one can prove that for any $n \geqslant 5$, the group $\mathfrak{A}_n$ is simple. See Exercise 4.16.

4.7 Matrix groups

Let k be a field. The group of $n \times n$ invertible matrices is denoted $\mathrm{GL}(n,k)$. It has three interesting subgroups: T, consisting of diagonal matrices; B, upper-triangular matrices; and U, the set of upper-triangular matrices with only 1s on the diagonal.

Proposition 4.7.1. *The group U is a nilpotent group. The derived subgroup of B is contained in U. In particular, B is a solvable group.*

Proof. Diagonal coefficients of the product of two matrices in B are just the product of their respective diagonal coefficients. This shows that any commutator of matrices in B belongs to U, whence $D(B) \subset U$. If we prove that U is nilpotent, then it is solvable, so $D(B)$ is solvable too (Proposition 4.5.2, b)). By Prop. 4.5.2, c), this implies that B is solvable, for $D(B)$ is a normal subgroup of B with $B/D(B)$ abelian (see Exercise 4.3).

Let us now show that U is nilpotent. Let us consider the canonical basis $(e_1, \dots, e_n)$ of the vector space $V = k^n$. Let $V_0 = 0$, and for each m, $1 \leqslant m \leqslant n$, let V_m be the vector space spanned by $(e_1, \dots, e_m)$. We also set $V_m = 0$ for $m \leqslant 0$. For any integer m, let N_m be the set of all endomorphisms u of V such that $u(V_i) \subset V_{i-m}$ for all i. For any $u \in N_m$ and $v \in N_p$, one has $u \circ v \in N_{m+p}$. Since $N_m = 0$ for $m \geqslant n$, this implies that for any $m \geqslant 1$, endomorphisms in N_m are *nilpotent*. Moreover, for all m, N_m is vector subspace of the space of all endomorphisms of V.

For any integer $m \geqslant 1$, let G_m denote the set of endomorphisms $u \in \operatorname{End}(V)$ such that $u - \mathrm{id} \in N_m$. If $u = \mathrm{id} + v$ and $u' = \mathrm{id} + v'$ belong to G_m, for some $m \geqslant 1$, then $uu' = \mathrm{id} + v + v' + vv'$ belongs to G_m, since $v + v' + vv' \in N_m$. Moreover, $vv' \in N_{2m} \subset N_{m+1}$. Let us also remark that for any nilpotent endomorphism $v \in \operatorname{End}(V)$, $v^n = 0$ and $\mathrm{id} + v$ is invertible, its inverse being given by the formula

$$(\mathrm{id} + v)^{-1} = \mathrm{id} - v + v^2 - \dots + (-1)^{n-1} v^{n-1}.$$

If $v \in N_m$, this shows that $(\mathrm{id} + v)^{-1}$ belongs to G_m. Consequently, G_m is a subgroup of $\mathrm{GL}(V)$.

Let $u \in G_m$ and let $g \in U$. If $v = u - \mathrm{id}$, one has

$$gug^{-1} = g(\mathrm{id} + v)g^{-1} = \mathrm{id} + gvg^{-1}.$$

Since $v \in N_m$, the endomorphism gvg^{-1} maps $g(V_i)$ into $g(V_{i-m})$ for any integer i ; since $g \in U$, $g(V_i) = V_i$ for any integer i, so that $gvg^{-1} \in N_m$. This shows that $gug^{-1} \in G_m$, hence that G_m is a normal subgroup in U.

The preceding calculations show moreover that for any $u = \mathrm{id} + v$ and $u' = \mathrm{id} + v' \in G_m$,

$$uu'u^{-1}(u')^{-1} = (\mathrm{id} + v + v' + w)(\mathrm{id} - v - v' + w') = \mathrm{id} + w'',$$

for some elements w, w' and $w'' \in N_{2m} \subset N_{m+1}$. In particular, G_m/G_{m+1} is contained in the center of U/G_{m+1}.

The chain of subgroups

$$\{1\} = G_n \subset G_{n-1} \subset \cdots \subset G_1 = U$$

allows us to conclude that U is a nilpotent group. □

The following converse is of fundamental importance for the algebraic theory of differential equations.

Theorem 4.7.2 (Lie, Kolchin). *Any solvable connected subgroup of the group* $\mathrm{GL}(n, \mathbf{C})$ *is conjugate to a subgroup of* B.

The proof uses the following classical lemma.

Lemma 4.7.3. *For any family of matrices which commute pairwise, there is a basis in which all the matrix are simultaneously in triangular form. In particular, any commutative subgroup of* $\mathrm{GL}(n, \mathbf{C})$ *is conjugate to a subgroup of* B.

Proof. Let us show the lemma by induction on n, the case $n = 1$ being trivial.

Let G be a family of elements of $n \times n$ matrices with complex coefficients, commuting pairwise. If G consists only of scalar elements (also called homotheties), that is, elements of the form $\lambda \mathrm{id}$ with $\lambda \in \mathbf{C}$, the result is clearly true. Otherwise, we may find a nonscalar element $h \in G$ and consider $V \subset \mathbf{C}^n$ one of its eigenspaces, with corresponding eigenvalue $\lambda \in \mathbf{C}$. (In particular, $d = \dim V \in \{1, \ldots, n-1\}$.) For any $g \in G$,

$$hg(v) = gh(v) = g(\lambda v) = \lambda g(v),$$

hence $g(v) \in V$. This shows that V is invariant under multiplication by any matrix in G. Let $(e_1, \ldots, e_d)$ be a basis of V and complete it to a basis $\mathscr{B} = (e_1, \ldots, e_n)$ of $\mathbf{C}^n$. In this basis $\mathscr{B}$, any matrix in G has a block representation $\left(\begin{smallmatrix} g_1 & * \\ 0 & g_2 \end{smallmatrix}\right)$ with $g_1 \in \mathrm{Mat}(d, \mathbf{C})$ and $g_2 \in \mathrm{Mat}(n-d, \mathbf{C})$. By the usual product of block-matrices, all such matrices g_1 commute with one another, as do the matrices g_2. By induction, there is a basis $f_1, \ldots, f_d$ of V in which all these matrices g_1 are upper-triangular, and similarly a basis $f_{d+1}, \ldots, f_n$ of $\mathrm{vect}(e_{d+1}, \ldots, e_n)$ in which all matrices g_2 are upper-triangular. Finally, in the basis $(f_1, \ldots, f_n)$, all matrices of G are upper-triangular, as was to be shown. □

Proof of Lie–Kolchin's theorem. The proof is by induction on the dimension n.

First assume that there is a subspace $V \subset \mathbf{C}^m$, with $V \neq 0$ and $V \neq \mathbf{C}^m$, which is left invariant by all elements of G. By taking a basis of V, a supplementary subspace W and a basis of W, we can assume that matrices in G are block-triangular: $\left(\begin{smallmatrix} g_1 & * \\ 0 & g_2 \end{smallmatrix}\right)$. The image of the group morphism $G \to \mathrm{GL}(V)$ defined by $g \mapsto g_1$ is a solvable subgroup of $\mathrm{GL}(V)$. By induction, there is a basis of G such that all matrices g_1 are upper-triangular. Similarly, the image of the group morphism $G \to \mathrm{GL}(W)$, $g \mapsto g_2$, is a solvable subgroup of $\mathrm{GL}(W)$ and W has a basis such that all such matrices g_2 are upper-triangular. These bases of V and W form a basis of $\mathbf{C}^n$ in which all matrices in G are upper-triangular.

We can therefore assume that no subspace of $\mathbf{C}^n$ is invariant under G, except 0 and $\mathbf{C}^n$. By induction on the least integer m with $D^m(G) = \{1\}$, let us now show that $n = 1$.

If $D^1(G) = \{1\}$, that is if G is abelian, Lemma 4.7.3 implies that there is an invertible matrix P such that $PGP^{-1} \subset B$. Since the line $\mathbf{C}e_1$ generated by the first basis vector e_1 of $\mathbf{C}^n$ is stable by any matrix in B, the line $\mathbf{C}(Pe_1)$ is stable by G, hence $n = 1$.

Now assume we have proved $n = 1$ if $D^{m-1}(G) = \{1\}$ and consider a subgroup $G \in \mathrm{GL}(n, \mathbf{C})$ with $D^m(G) = \{1\}$ but $D^{m-1}(G) \neq \{1\}$. In particular, $H = D^{m-1}(G)$ satisfies $D(H) = \{1\}$ and there is $P \in \mathrm{GL}(n, \mathbf{C})$ such that $PHP^{-1} \subset B$. Replacing G by PGP^{-1}, we assume that $H \subset B$.

In particular, e_1 is an eigenvector of any $h \in H$. Let then $V \subset \mathbf{C}^n$ be the subspace generated by all $v \in \mathbf{C}^n$ which are eigenvectors of any $h \in H$. It contains e_1, so $V \neq \{0\}$. Let us show that it is invariant under G. By linearity, it is sufficient to show that for any $g \in G$ and any $v \in V$ which is an eigenvector of any $h \in H$, then $g(v) \in V$. Indeed, if $h \in H$, then

$$h(g(v)) = gg^{-1}hg(v) = g(g^{-1}hg(v)).$$

Since $g^{-1}hg \in H$, there is $\lambda \in \mathbf{C}$ such that $g^{-1}hg(v) = \lambda v$. It follows that $h(g(v)) = \lambda g(v)$ and $g(v)$ is an eigenvector of $h \in H$. Therefore $g(v) \in V$ and V is invariant under g. The first reduction implies that $V = \mathbf{C}^n$, so that $\mathbf{C}^n$ has a basis consisting of eigenvectors of any element of H. In that basis, any matrix in H is diagonal. A new change of basis allows us to assume that $H \subset T$.

Let us now fix $h \in H$, $h \neq \{1\}$. Since $H = D^{m-1}(G)$ is a normal subgroup of G, one has $g^{-1}hg \in H$ for any $g \in G$ and $h \in H$. Since the map $G \to H$ given by $g \mapsto g^{-1}hg$ is continuous and since G is connected, its image is connected in H. But for $g \in G$, $g^{-1}hg$ is a diagonal matrix with the same eigenvalues as h, and there are only finitely many such matrices. Since a connected finite set in $\mathrm{GL}(n, \mathbf{C})$ has only one element, the map $g \mapsto g^{-1}hg$ is constant, with value $h = e^{-1}he$. Consequently, for any $g \in G$ and any $h \in H$, $gh = hg$. In other words, $H = D^{m-1}(G)$ is contained in the center of G.

Let W be any eigenspace for h. For any $g \in G$, we proved that g and h commute, so W is invariant by g. This shows that W is stable by G. By the first reduction, this implies that $W = \mathbf{C}^n$. For any $h \in H$ there thus exists $\lambda_h \in \mathbf{C}^*$ such that $h = \lambda_h \mathrm{id}$.

The determinant of a commutator is equal to 1. Since $m - 1 \geqslant 1$, $H \subset \mathrm{SL}(n, \mathbf{C})$ and $\lambda_h^n = 1$ for any $h \in H$. This shows that H is a finite group. By the following lemma, H is connected, whence $H = \{1\}$, our desired contradiction. □

Lemma 4.7.4. *Let G be a connected subgroup of* $\mathrm{GL}(n, \mathbf{C})$. *Then its derived subgroup is also connected.*

Proof. The set S of all commutators in G is the image of $G \times G$ by the continuous map $(g_1, g_2) \mapsto g_1 g_2 g_1^{-1} g_2^{-1}$. Therefore, S is connected.

Let S_m be the set of all products $s_1 \dots s_m$, with $s_i \in S$ and $m \geqslant 1$. It is the image of S^m under the continuous map $(g_1, \dots, g_m) \mapsto g_1 \dots g_m$. Since S is connected, so are S^m and S_m.

Since the inverse of a commutator is again a commutator, one has $D(G) = \{e\} \cup \bigcup_{m \geqslant 1} S_m$. Since the S_m have the identity matrix in common, $D(G)$ is connected, q.e.d. □

Exercises

Exercise 4.1. **a)** Let m and n be two coprime integers. Show that $(\mathbf{Z}/mn\mathbf{Z})^*$ is isomorphic to $(\mathbf{Z}/m\mathbf{Z})^* \times (\mathbf{Z}/n\mathbf{Z})^*$.

b) If $(\mathbf{Z}/n\mathbf{Z})^*$ is a cyclic group, show that there exists an odd prime number p and an integer $m \geqslant 0$ such that $n = p^m$, or $n = 2p^m$, or $n = 4$.

c) Using Exercise 3.15, show the converse of *b*).

Exercise 4.2. Recall that $\mathbf{H}$ denotes the (noncommutative) field of quaternions.

a) Show that the polynomial equation $X^2 = 1$ has infinitely many solutions in $\mathbf{H}$.

Let G be the subset $\{\pm 1, \pm i, \pm j, \pm k\}$ in $\mathbf{H}^*$.

b) Show that G is a subgroup of $\mathbf{H}^*$. Show that it is not commutative. It particular, it cannot be cyclic, which shows that the conclusion of Exercise 1.16 does not hold anymore without the assumption that the field is commutative.

c) Show that any subgroup of G is normal.

Exercise 4.3. Let G be a group.

a) Show that the quotient of G by its derived subgroup $D(G)$ is an abelian group.

b) Let H be any subgroup of G. Show that H is the kernel of a homomorphism from G to an abelian group if and only if H contains $D(G)$. (You might want to prove *first* that a subgroup containing the derived subgroup is normal.)

Exercise 4.4. **a)** Let G be a group, and let A and B be two normal subgroups of G such that $A \cap B = \{e\}$. If $a \in A$ and $b \in B$, show that $aba^{-1}b^{-1} = e$, hence that a and b commute. Conclude that the subgroup $AB \subset G$ generated by A and B is isomorphic to the direct product $A \times B$.

b) Let G be a finite group, $(A_i)_{1 \leqslant i \leqslant r}$ a family of normal subgroups of G. Let $n_i =$ card A_i; assume that the integers n_i are pairwise coprime, and that $\prod_{i=1}^{r} n_i = \operatorname{card} G$. Show that G is isomorphic to the direct product $\prod_{i=1}^{r} A_i$.

Exercise 4.5 (Semidirect product). **a)** Let A and B be two groups, and let $\varphi \colon B \to \operatorname{Aut}(A)$ a morphism of groups (in other words, φ defines an action of B on A by group automorphisms).

Endow the product set $G = A \times B$ by the composition defined by

$$(a, b) \cdot (a', b') = (a\varphi(b)(a'), bb').$$

Show that it is a group law. The group G with this law is called the semi-direct product of A and B, and usually denoted $A \rtimes_\varphi B$.

b) Show that the map $G \to B$ given by $(a, b) \mapsto b$ is a surjective morphism of groups. Show that its kernel is isomorphic to A. Show that the map $B \to G$ given by $b \mapsto (\varphi(b)(e), b)$ is a morphism of groups.

c) For any $g = (a, b) \in G$, define $\sigma(g) \colon A \to A$ by $\sigma(g)(x) = a\varphi(b)(x)$. Show that $\sigma(g)$ is a permutation of A, and that the map $\sigma \colon G \to \mathfrak{S}(A)$ is an injective morphism of groups.

d) Let G be a group, A a normal subgroup in G, and let $\pi \colon G \to B = G/A$ be the canonical projection. Assume that there is a group morphism $f \colon B \to G$ such that $\pi(f(b)) = b$ for any $b \in B$ (one says that f is a *section* of π). Show that there is a morphism of groups $\varphi \colon B \to \operatorname{Aut}(A)$ such that $\varphi(b)(a) = f(b)af(b^{-1})$ for any $a \in A$ and any $b \in B$. Then show that G is isomorphic to $A \rtimes_\varphi B$.

Exercise 4.6. Fix an integer $n \geqslant 2$, and let $\zeta = \exp(2i\pi/n)$. Consider the standard regular n-gon Γ in the plane $\mathbf{R}^2$ with vertices the points A_k of coordinates $(\cos(2k\pi/n), \sin(2k\pi/n))$, $1 \leqslant k \leqslant n$. Let O denote the origin of the plane.

Let D_n be the group of affine transformations of the plane that preserve the vertices of Γ.

a) Show that D_n contains the cyclic group of order n generated by the rotation r of center O and angle $2\pi/n$. Show that it contains also the orthogonal symmetry s with respect to axis Ox.

b) Let $g \in D_n$. Show that $g(O) = O$ and that $\det(g) \in \{\pm 1\}$. If moreover $\det(g) = 1$ and g fixes one vertex of Γ, then $g = \mathrm{id}$. Conclude that D_n is generated by r and s.

c) Show that $srs = r^{-1}$ and that D_n is isomorphic to the semidirect product $(\mathbf{Z}/n\mathbf{Z}) \rtimes_\varphi \{\pm 1\}$, where $\varphi \colon \{\pm 1\} \to \operatorname{Aut}(\mathbf{Z}/n\mathbf{Z})$ is the map defined by $\varphi(\varepsilon)(m) = \varepsilon m$.

Exercise 4.7. Let G be a finite group; assume that the cardinality of G is the power of a prime number p.

a) Let X be a finite set on which G acts. Let X^G be the set of fixed points of G, that is, the set of all $x \in X$ such that $g \cdot x = x$ for any $g \in G$. Show that

$$\operatorname{card}(X) \equiv \operatorname{card}(X^G) \pmod{p}.$$

b) Let Z be the center of G. Applying the class formula to the conjugation action of G on itself, show that $Z \neq \{1\}$.

c) Let G be a finite group with p^2 elements. Show that G is commutative.

Exercise 4.8 (Sylow's first theorem). Let G be a finite group, p a prime number. Let p^r the highest power of p that divides the order of G.

a) Let X be the family of all subsets of G having p^r elements. Show that the cardinality of X is not divisible by p.

b) Consider the action of G on X by left translations: if $A \subset G$, $g \cdot A = \{ga\,;\, a \in A\}$. Show that the stabilizer $\{g \in G\,;\, g \cdot A = A\}$ of any $A \in X$ is a subgroup of G with cardinality $\leqslant p^r$.

c) Show that there exists $A \in X$ whose orbit has cardinality prime to p. Using the class formula, prove that the stabilizer of A is a subgroup of G with exactly p^r elements (*p-Sylow subgroup*).

d) More generally, show as follows that for any $s \leqslant r$, G contains subgroups with p^s elements. Let X be the family of subsets of G with p^s elements. Prove that $\operatorname{ord}_p(\operatorname{card} X) = r - s$, where $\operatorname{ord}_p(m)$ denotes the highest power of p that divides the integer m. Show that there exists $A \in X$ such that its orbit $\mathscr{O}(A)$ satisfies $\operatorname{ord}_p(\operatorname{card} \mathscr{O}(A)) \leqslant r - s$. Show that $\operatorname{ord}_p(\operatorname{card} \operatorname{Stab} A) \geqslant s$ and conclude that $\operatorname{Stab} A$ is a subgroup of G with p^s elements.

Exercise 4.9 (Sylow's second theorem). Let G be a finite group, p a prime number. Let p^r be the highest power of p that divides the order of G. Let P be a p-Sylow subgroup of G, that is a subgroup with cardinality p^r.

a) Let $H \subset G$ be a subgroup whose order is a power of p. Introduce the action of H by left translations on the set G/P of right cosets modulo P, in other words, the action defined by $h \cdot gP = (hg)P$. Show that it has a fixed point: there exists $g \in G$ such that for any $h \in H$, $hgP = gP$ (use the class formula). Conclude that $H \subset gPg^{-1}$.

b) Show that any two p-Sylow subgroups are conjugate.

c) Let N denote the normalizer of P in G, i.e. , the set of all $g \in G$ such that $gPg^{-1} = P$. Show that $\operatorname{card}(N/P) \equiv \operatorname{card}(G/P) \pmod{p}$. Show that the number of p-Sylow subgroups in G divides $\operatorname{card}(G)/p^r$ and is congruent to 1 modulo p.

Exercise 4.10. Let G be a group.

a) Set $D^0 = G$, and, for each i, define D^{i+1} to be the derived subgroup of D^i. Show that G is solvable if and only if there exists n such that $D^n = \{1\}$.

b) Set $C^0 = G$ and if $i \geqslant 0$, define C^{i+1} to be the subgroup of G generated by all commutators $ghg^{-1}h^{-1}$ with $g \in G$ and $h \in C^i$ (*descending central series*). Show that G is nilpotent if and only if there is n with $C^n = \{1\}$.

Exercise 4.11. **a)** Show that a subgroup or a quotient of a nilpotent group is again nilpotent.

b) Let G be a group, H a subgroup of G which is contained in the center of G. Observe that H is normal in G. If G/H is nilpotent, show that G is nilpotent.

c) Show that a finite group the cardinality of which is a power of a prime number is nilpotent. (Use Exercise 4.7, *b*).)

Exercise 4.12. **a)** Let G be a finite group, H a normal subgroup of G, and let π denote the canonical surjective morphism $G \to G/H$. Let p be a prime number and P a p-Sylow subgroup in G. Show that $P \cap H$ is a p-Sylow subgroup in H, and that $\pi(P)$ is a p-Sylow subgroup in G/H.

b) Let G be a finite group, $C \subset G$ a subgroup contained in the center of G, and let P, P' be two p-Sylow subgroups of G such that $P \cap C = P' \cap C$ and $PC = P'C$. For any $g \in P$, show that there are $\pi(g) \in P'$ and $c(g) \in C$ such that $g = \pi(g)c(g)$. Show that the map $g \mapsto c(p)(P' \cap C)$ from P to $C/(P' \cap C)$ is well defined and is a morphism of groups. Show that it maps any element to the identity. Conclude that $P = P'$.

c) Let G be a finite nilpotent group. Show by induction on the cardinality of G that G has a single p-Sylow subgroup.

d) Let G be a finite nilpotent group. Show that G is isomorphic to the direct product of its p-Sylow subgroups, for p dividing the order of G. (*Use Exercise 4.4.*)

e) Conversely, show that any finite group which is the direct product of its Sylow subgroups is a nilpotent group.

Exercise 4.13. This exercise proposes a proof of a famous theorem of Wedderburn: *Any finite field is commutative.* Thus let F be a finite field which is not assumed to be commutative.

a) Let Z be the center of F, that is the set of all $a \in F$ such that $ax = xa$ for any $x \in F$. Show that Z is a commutative subfield in F. Let q be its cardinality. Show that there exists an integer $n \geqslant 1$ such that $\operatorname{card} F = q^n$.

b) Let $x \in F$. Show that the set C_x consisting of all $a \in F$ such that $ax = xa$ is a subfield in F. Show that there is a positive integer n_x dividing n such that $\operatorname{card} C_x = q^{n_x}$. (Notice that multiplication on the left by elements of C_x gives F the structure of a C_x-vector space.)

c) For $x \in F^*$, compute in terms of n_x the cardinality of its conjugacy class $\mathscr{C}(x)$ (the set of all $axa^{-1} \in F$, for $a \in F^*$).

d) Assuming $x \notin Z$, show that the cardinality of $\mathscr{C}(x)$ is a multiple of $\Phi_n(q)$. (Φ_n is the nth cyclotomic polynomial.)

e) Using the class formula, show that $q^n - q$ is divisible by $\Phi_n(q)$. Conclude that $n = 1$, hence that F is commutative.

Exercise 4.14. Let G be a transitive subgroup of $\mathfrak{S}_n$. For $i \in \{1, \dots, n\}$, denote by G_i the set of all $g \in G$ such that $g(i) = i$.

a) Show that G_i is a subgroup of G. Show that $(G : G_i) = n$.

b) Show that $\bigcup_{i=1}^{n} G_i \neq G$. (Bound from above the cardinality of the left-hand side.) Conclude that there is an element of G without any fixed point (Jordan, 1872).

Exercise 4.15. Let G be a finite group acting on a finite set X. For any $g \in G$, denote by $f(g)$ the number of fixed points of g.

a) Show Burnside's formula: There are exactly

$$\frac{1}{\operatorname{card} G} \sum_{g \in G} f(g)$$

orbits of G in X. (Count in two ways the number of elements in $G \times X$ such that $g \cdot x = x$, first summing over $g \in G$, then summing over $x \in X$.)

b) Show that

$$\frac{1}{\operatorname{card} G} \sum_{g \in G} f(g)^2 \geqslant 2.$$

(Look at the action of G on $X \times X$.)

c) Assume that the action of G on X is transitive. Show by summing over $g \in G$ the quantity $(f(g) - 1)(\operatorname{card} X - f(g))$ that at least $\operatorname{card} G / \operatorname{card} X$ elements of G have no fixed point in X. (This improvement of Exercise 4.14 is due to Cameron and Cohen.)

Exercise 4.16 (Simplicity of $\mathfrak{A}_n$ for $n \geqslant 5$). In this exercise, you will prove that the alternating group $\mathfrak{A}_n$ is a simple group, provided $n \geqslant 5$.

a) Let N be a normal subgroup of $\mathfrak{A}_5$. Show as follows that N contains a 3-cycle:

1) First assume that N contains a double transposition, say $\sigma = (1,2)(3,4)$. Show that the permutation $\tau = (1,5)(3,4)$ is conjugate to σ in $\mathfrak{A}_n$. Notice that $\sigma\tau$ is a 3-cycle in N.

2) In the case where N contains a permutation of order 5, say $\sigma = (1,2,3,4,5)$, show that $\tau = (2,3,1,4,5)$ is conjugate to σ and conclude that N contains the 3-cycle $(4,1,2)$.

b) Show that $\mathfrak{A}_5$ is simple.

c) Let $n \geqslant 6$ and assume by induction that $\mathfrak{A}_{n-1}$ is simple. We want to show that $\mathfrak{A}_n$ is simple. Let N be a normal subgroup of $\mathfrak{A}_n$, with $N \neq \{1\}$ and $N \neq \mathfrak{A}_n$.

Let $\sigma \in N$, $\sigma \neq \mathrm{id}$; show that $n \neq \sigma(n)$.

d) (*continued*) Let $\sigma \in N \setminus \{\mathrm{id}\}$. Choose two integers $i \neq j$, distinct from n and $\sigma(n)$ and consider $\tau = (i,j)(n, \sigma(n))$. Show that $\sigma' = \sigma\tau\sigma\tau^{-1} \in N$ but that $\sigma'(n) = n$. Using the induction hypothesis, conclude that $\sigma' = \mathrm{id}$.

e) (*continued*) Considering the equality $\tau\sigma\tau^{-1} = \sigma$, show that $\sigma^2(n) = n$ and that

$$\{\sigma(i), \sigma(j)\} = \{i, j\}.$$

Conclude that $\sigma = (n, \sigma(n))$, which contradicts the fact that $\sigma \in \mathfrak{A}_n$. Hence N does not exist and $\mathfrak{A}_n$ is simple.

Exercise 4.17. Let n be any integer, with $n \geqslant 5$, and let G be a subgroup of the symmetric group $\mathfrak{S}_n$. Let $d = (\mathfrak{S}_n : G)$ be the index of G in $\mathfrak{S}_n$.

a) Assuming that G is a normal subgroup of $\mathfrak{S}_n$, show that $G = \mathfrak{A}_n$ or $G = \mathfrak{S}_n$. (Recall from Exercise 4.16 that $\mathfrak{A}_n$ is simple.)

b) Show that there is a morphism of groups $\mathfrak{S}_n \to \mathfrak{S}_d$ whose kernel is contained in G.

c) If $G \neq \mathfrak{A}_n$ and $G \neq \mathfrak{S}_n$, show that $d \geqslant n$.

5

Applications

We see in this chapter how Galois theory can be used to get a satisfactory answer to the problem of constructions with ruler and compass. By analogous methods, we discuss the problem of solving polynomial equations using radicals and we show how Galois theory allows us to understand the explicit resolution of equations of degrees up to 4. Finally, we will study the behavior of the Galois group of an equation when we vary the coefficients.

5.1 Constructibility with ruler and compass

Let us go back to the problem of geometric constructions with ruler and compass. We are mostly interested here in complex numbers which are constructible from the set $\{0,1\}$. By Wantzel's theorem (Theorem 1.4.1), these are the complex numbers z for which there is a sequence of extensions, $\mathbf{Q} = K_0 \subset K_1 \subset K_2 \subset \cdots \subset K_n$, such that $z \in K_n$ and such that for any i, $[K_i : K_{i-1}] = 2$. The main result is the following.

Theorem 5.1.1. *An algebraic number $z \in \mathbf{C}$ is constructible (from $\{0,1\}$) if and only if the degree of the extension of $\mathbf{Q}$ generated by z and its conjugates is a power of 2.*

To understand step by step what happens, let us begin by proving the following proposition.

Proposition 5.1.2. *Let $z \in \mathbf{C}$ be a constructible number. Then any conjugate of z is constructible.*

Proof. Let $\mathbf{Q} = K_0 \subset K_1 \subset \cdots \subset K_n$ be a sequence of quadratic extensions such that $z \in K_n$. Let $\mathbf{Q} \subset L$ be a Galois extension such that $K_n \subset L$. If z' is a conjugate of z, there exists an element $\sigma \in \mathrm{Gal}(L/\mathbf{Q})$ such that

$\sigma(z) = z'$. (This is essentially the content of the proof of Proposition 3.3.2; see Exercise 3.6.) Set $K'_j = \sigma(K_j)$ for $0 \leqslant j \leqslant n$. These are subfields of L and for any j, $K'_{j-1} \subset K'_j$, with $[K'_j : K'_{j-1}] = 2$. Since $z' \in K'_n$, this shows that z' is constructible. □

Proof of Theorem 5.1.1. Now let $z \in \mathbf{C}$ be a constructible number and let L be the extension of $\mathbf{Q}$ generated by the conjugates of z. By Theorem 1.1.3, any element in L is constructible. But $\mathbf{Q}$ having characteristic zero, it follows from the Primitive Element Theorem (Theorem 3.3.3) that there is $\alpha \in L$ such that $L = \mathbf{Q}[\alpha]$. This element α is constructible, so its degree is a power of 2, by Corollary 1.4.4. It follows that $[L : \mathbf{Q}]$ is a power of 2, which was to be shown.

Conversely, assume that $[L : \mathbf{Q}]$ is a power of 2. Since L is generated by the roots of the minimal polynomial of z, it is a splitting extension of a separable polynomial ($\mathbf{Q}$ has characteristic zero), hence a Galois extension (Proposition 3.2.7). The order of its Galois group $G = \mathrm{Gal}(L/K)$ is a power of 2. By Lemma 5.1.3 below, applied to $p = 2$, there exist subgroups $\{1\} = G_0 \subset G_1 \subset \cdots \subset G_n = G$, each of index 2 in the next. They correspond to a sequence of extensions of $\mathbf{Q}$ contained in L, $\mathbf{Q} = L^G \subset L^{G_{n-1}} \subset \cdots \subset L^{G_0} = L$, with $[L^{G_j} : L^{G_{j+1}}] = (G_{j+1} : G_j) = 2$. By Wantzel's theorem 1.4.1, any element of L is then constructible. In particular, z is constructible. □

Lemma 5.1.3. *Let G be a finite group, the order of which is a power of a prime number p. Then G has a normal series*

$$\{1\} = G_0 \subset G_1 \subset \cdots \subset G_n = G$$

such that for any j, $(G_j : G_{j-1}) = p$.

Proof. We will argue by induction on the order of G. By Exercise 4.7, the center Z of G is a nontrivial commutative group. Let $g \in Z \setminus \{e\}$; the order of g divides $\operatorname{card} Z$, hence is a power of p, say p^a, with $a \geqslant 1$. It follows that $h = g^{p^{a-1}}$ is an element of Z of order p. Let G_1 denote the subgroup of G generated by h. It is a normal subgroup of order p in G. The cardinality of the group G/G_1 is a power of p, say p^m. By induction, there are subgroups $H_j \subset G/G_1$, for $0 \leqslant j \leqslant m$, such that H_{j-1} is a normal subgroup of H_j and $(H_j : H_{j-1}) = p$ for each j. For $2 \leqslant j \leqslant m+1$, let G_j denote the preimage of H_{j-1} in G/G_1. One has $G_1 \subset G_2 \subset \cdots \subset G_{m+1}$, G_{j-1} is a normal subgroup of G_j and $(G_j : G_{j-1}) = p$ for any $j \leqslant m+1$, and $G_{m+1} = G$. □

5.2 Cyclotomy

This name is the concatenation of two Greek roots, and it roughly means "cutting the circle." Consider a regular n-gon inscribed in the unit circle.

Its vertices divide the unit circle into n equal parts. By identifying points of the plane with the complex numbers, and assuming that one of the vertices is 1, these vertices correspond to nth roots of unity. Therefore, cyclotomy characterizes nowadays any study of mathematics that is related to roots of unity. For example, cyclotomic fields are fields generated by a root of unity, and the roots of the nth cyclotomic polynomial are exactly the primitive nth roots of unity.

We now obey the title of this section and begin by studying the Galois-theoretical properties of the equation $X^n = 1$.

Theorem 5.2.1. *Let K be a field, and let n be any positive integer. We assume that the characteristic of K does not divide n. Let $K \subset L$ be a splitting extension of the polynomial $X^n - 1$. It is a Galois extension, and its Galois group is isomorphic to a subgroup of the group $(\mathbf{Z}/n\mathbf{Z})^*$.*

More precisely, there is a canonical injective morphism of groups

$$\varphi\colon \operatorname{Gal}(L/K) \to (\mathbf{Z}/n\mathbf{Z})^*$$

such that for any nth root of unity $\zeta \subset L$ and any $\sigma \in \operatorname{Gal}(L/K)$,

$$\sigma(\zeta) = \zeta^{\varphi(\sigma)}.$$

Proof. Fix a primitive nth root of unity ζ. Since the polynomial $X^n - 1$ is separable, the extension $K \subset L$ is Galois. The roots of $X^n - 1$ are the ζ^m, for $0 \leqslant m \leqslant n-1$, hence $L = K(\zeta)$.

Let $\sigma \in \operatorname{Gal}(L/K)$; it maps ζ to a nth root of unity, which is of the form ζ^m for some integer m whose class modulo m is well defined. Moreover, if $\sigma(\zeta)^k = 1$, one has $\zeta^k = 1$, hence $\sigma(\zeta)$ is still a primitive root, so that m is prime to n. This defines a map $\varphi\colon \operatorname{Gal}(K(\zeta)/K) \to (\mathbf{Z}/n\mathbf{Z})^*$.

Let θ be any nth root of unity, and fix an integer a such that $\theta = \zeta^a$. One has

$$\sigma(\theta) = \sigma(\zeta^a) = \sigma(\zeta)^a = (\zeta^m)^a = \zeta^{ma} = \theta^m,$$

and $\sigma(\theta) = \theta^{\varphi(\sigma)}$. This shows in particular that the map φ does not depend on the choice of a particular primitive root ζ.

If $\sigma, \tau \in \operatorname{Gal}(\mathbf{Q}(\zeta)/\mathbf{Q})$, with $\sigma(\zeta) = \zeta^m$ and $\tau(\zeta) = \zeta^n$, one has

$$(\sigma \circ \tau)(\zeta) = \sigma(\zeta^n) = \sigma(\zeta)^n = \zeta^{mn},$$

so that $\varphi(\sigma \circ \tau) = \varphi(\sigma)\varphi(\tau)$. This implies that φ is a morphism of groups. Moreover, if $\varphi(\sigma) = 1$, then $\sigma(\zeta) = \zeta$. Since ζ generates $\mathbf{Q}(\zeta)$, this implies $\sigma = \operatorname{id}$ and φ is injective. □

We saw in Chapter 1, Example 1.4.7, that it is impossible to construct a regular 9-gon with ruler and compass. However, C.-F. Gauss had shown that the regular polygon with 17 edges is actually constructible (as he wrote in his mathematical diary, March 30, 1796). He was barely 19 years old. We now prove a general result about the possibility of constructing regular polygons with ruler and compass.

Theorem 5.2.2. *A regular polygon with n sides is constructible with ruler and compass if and only if n is the product of a power of 2 and of distinct Fermat primes.*

Recall that a *Fermat prime* is a prime number of the form $F_m = 2^{2^m}+1$, where m is an integer. Among them are 3, 5, 17, 257 and 65537, corresponding to $m = 0, \dots, 4$. Fermat had conjectured that all F_m's are prime numbers but Euler showed that 641 divides F_5. (*Exercise*: prove it; show also that if n is not a power of 2, then 2^n+1 is not a prime number.) Actually, the five Fermat primes just listed above are the only known ones! It has also been proved that $F_6, \dots, F_{16}$ are not primes.

Proof. Let $\mathscr{P}$ be the set of integers $n \geqslant 3$ such that one can construct a regular n-gon with ruler and compass. In other words, an integer $n \geqslant 3$ belongs to $\mathscr{P}$ if and only if the algebraic number $\exp(2i\pi/n)$ is constructible. Its conjugates are among nth root of unity.

Using the following remarks, however, we reduce ourselves to the case where n is a prime or the square of a prime.

a) If $n \in \mathscr{P}$, then $2n \in \mathscr{P}$.

Indeed, if a regular n-gon is already drawn, one just needs, for each edge AB of it, to draw the perpendicular to AB passing through the center O of the n-gon, for it cuts the angle $\widehat{AOB}$ into two equal parts.

b) If $n \in \mathscr{P}$, then any integer $m \geqslant 3$ dividing n also belongs to $\mathscr{P}$.

To construct a regular m-gon, just join every (n/m)th vertex of a regular n-gon.

c) If m and n are two coprime integers belonging to $\mathscr{P}$, then their product mn belongs to $\mathscr{P}$.

That m and n belong to $\mathscr{P}$ means that the two complex numbers $\exp(2i\pi/m)$ and $\exp(2i\pi/n)$ are constructible. Since m and n are coprime, there are integers u and v such that $um+vn=1$, hence

$$\exp(2i\pi/mn) = \exp\left(2i\pi\left(\frac{u}{n}+\frac{v}{m}\right)\right) = \left(\exp(2i\pi/n)\right)^u \left(\exp(2i\pi/m)\right)^v$$

is constructible, which in turns means that $mn \in \mathscr{P}$.

To prove the theorem, we now just need to prove that the only prime numbers in $\mathscr{P}$ are Fermat primes, and that $\mathscr{P}$ does not contain the square of any odd prime number. By Theorem 5.1.1, these two statements reduce to the following facts, where p is an odd prime number.

d) The complex number $\exp(2i\pi/p)$ *is an algebraic number of degree* $p-1$ *over* $\mathbf{Q}$. *The extension of* $\mathbf{Q}$ *generated by all pth roots of unity has degree* $p-1$.

Let P be the minimal polynomial of $\exp(2i\pi/p)$. It is a monic polynomial with integer coefficients and it divides $(X^p-1)/(X-1) = 1+X+\cdots+X^{p-1}$, hence there is $Q \in \mathbf{Z}[X]$ with

$$\frac{X^p - 1}{X - 1} = P(X)Q(X).$$

Set $a = \deg P$, $b = \deg Q$; in particular, $a+b = p-1$. Since $\exp(2i\pi/p)$ is not a rational number, $a \geqslant 2$.

Modulo p, one has $X^p - 1 \equiv (X-1)^p$. By uniqueness of decomposition into irreducible factors over $\mathbf{Z}/p\mathbf{Z}$, there are polynomials A and $B \in \mathbf{Z}[X]$ such that $P = (X-1)^a + pA(X)$, $Q = (X-1)^b + pB(X)$, Consequently,

$$\begin{aligned}\frac{X^p - 1}{X - 1} &= P(X)Q(X) \\ &= (X-1)^{a+b} + p\bigl(A(X)(X-1)^b + B(X)(X-1)^a\bigr) + p^2 A(X)B(X).\end{aligned}$$

Now evaluate the two sides of this equality at 1. If one had $b \geqslant 1$, it would follow that $p = p^2 AB(1)$, which is obviously a contradiction, for $AB(1)$ is an integer. Hence, $b = 0$, and $a = p-1$.

The last assertion comes from the fact that $\exp(2i\pi/p)$ generates the splitting extension over $\mathbf{Q}$ of the polynomial $X^p - 1$.

e) The complex number $\exp(2i\pi/p^2)$ *is an algebraic number of degree* $p(p-1)$ *over* $\mathbf{Q}$.

We do a similar analysis with the polynomial $X^{p^2} - 1$ divided by its factor $X^p - 1$, which does not vanish at $\exp(2i\pi/p^2)$. If $P \in \mathbf{Z}[X]$ is the minimal polynomial of $\exp(2i\pi/p^2)$, there is as above a polynomial $Q \in \mathbf{Z}[X]$ such that

$$\frac{X^{p^2} - 1}{X^p - 1} = P(X)Q(X).$$

Since $X^{p^2} - 1 = (X-1)^{p^2}$ modulo p, we similarly find polynomials A and $B \in \mathbf{Z}[X]$ with $P = (X-1)^a + pA$ and $Q = (X-1)^b + pB$, where $a = \deg P \geqslant 2$ and $b = \deg Q$. Evaluating the resulting equality

$$\frac{X^{p^2} - 1}{X^p - 1} = (X-1)^{p^2-p} + p\bigl((X-1)^a B(X) + (X-1)^b A(X)\bigr) + p^2 A(X)B(X)$$

at 1, we find as above that $b = 0$, hence the degree of $\exp(2i\pi/p^2)$ equals $a = p^2 - p$. □

Remark 5.2.3. These last two statements *d*) and *e*) are particular cases of a general theorem of Gauss, according to which the degree of $\exp(2i\pi/n)$ is equal to Euler's totient function $\varphi(n)$ (see Exercise 2.5). Together with Theorem 5.2.1, this shows that the Galois group of the extension $\mathbf{Q} \subset \mathbf{Q}(\exp(2i\pi/n))$ is isomorphic to $(\mathbf{Z}/n\mathbf{Z})^*$.

These particular cases, where $n = p^r$ is a power of a prime p, are usually proved using Eisenstein's criterion (Exercise 1.10). Indeed, applied to the polynomial $\Phi_{p^r}(Y+1)$ and the prime p, this criterion allows one to show that Φ_{p^r} is irreducible.

Corollary 5.2.4 (Gauss, 1801). *The regular polygon with* 17 *vertices is constructible with ruler and compass.*

Let us explain Gauss's explicit resolution of the equation $X^{17} = 1$. Let ζ be a primitive 17th root of unity in $\mathbf{C}$. The extension $\mathbf{Q} \subset \mathbf{Q}(\zeta)$ is Galois and its Galois group is isomorphic to $(\mathbf{Z}/17\mathbf{Z})^*$. Gauss's fundamental remark is that this group is cyclic, generated, for example, by the class of 3. Its powers modulo 17 are successively

$$1, 3, 9, 10, 13, 5, 15, 11, 16, 14, 8, 7, 4, 12, 2, 6, 1 \dots$$

Let $\sigma \in \operatorname{Gal}(\mathbf{Q}(\zeta)/\mathbf{Q})$ be the corresponding generator, mapping ζ to ζ^3, and set

$$a_0 = \sum_{k=0}^{7} \sigma^{2k}(\zeta) = \zeta + \zeta^9 + \zeta^{13} + \zeta^{15} + \zeta^{16} + \zeta^8 + \zeta^4 + \zeta^2,$$

$$a_1 = \sum_{k=0}^{7} \sigma^{2k+1}(\zeta) = \zeta^3 + \zeta^{10} + \zeta^5 + \zeta^{11} + \zeta^{14} + \zeta^7 + \zeta^{12} + \zeta^6.$$

One has $\sigma(a_0) = a_1$ and $\sigma(a_1) = a_0$. It follows that a_0 and a_1 are the two roots of a quadratic equation in $\mathbf{Q}[X]$. Precisely, one has $a_0 + a_1 = -1$ and $a_0 a_1 = -4$, so that

$$a_0,\ a_1 = \frac{-1 \pm \sqrt{17}}{2}.$$

The choice of signs depends on the choice of ζ. If $\zeta = \exp(2i\pi/17)$, a numerical calculation shows that $a_0 = (-1 + \sqrt{17})/2$. Set $K_1 = \mathbf{Q}(\sqrt{17})$. The Galois group of the extension $K_1 \subset \mathbf{Q}(\zeta)$ is generated by σ^2.

We continue by defining, for $0 \leqslant i \leqslant 3$,

$$b_i = \sum_{k=0}^{3} \sigma^{4k+i}(\zeta),$$

so that $\sigma(b_i) = b_{i+1}$ if $i = 0, 1, 2$ and $\sigma(b_3) = b_1$. In particular, b_0 and b_2 are permuted by σ^2, they are the two roots of a quadratic equation in K_1. One has $b_0 + b_2 = a_0$ and $b_0 b_2 = -1$, so that

$$b_0,\ b_2 = \frac{1}{2}\left(a_0 \pm \sqrt{a_0^2 + 4}\right) = -\frac{1}{4} + \frac{1}{4}\sqrt{17} \pm \sqrt{34 - 2\sqrt{17}},$$

and again, choosing the positive square root for a positive real number, numerical calculations show that b_0 is given by the formula with the + sign. Similarly,

$$b_1,\ b_2 = -\frac{1}{4} - \frac{1}{4}\sqrt{17} \pm \sqrt{34 + 2\sqrt{17}}.$$

Set $K_2 = \mathbf{Q}(\sqrt{34 - 2\sqrt{17}})$. The extension $K_2 \subset \mathbf{Q}(\zeta)$ is Galois, with group generated by σ^4.

Now define, for $0 \leqslant i \leqslant 7$,

$$c_i = \sigma^i(\zeta) + \sigma^{i+8}(\zeta).$$

The quantities c_0 and c_4 are permuted under σ^4, hence are the two roots of a quadratic equation over K_2. Concretely, $c_0 + c_4 = a_0$ and $c_0 c_4 = b_1$, hence

$$c_0,\ c_4 = \frac{1}{2}\left(a_0 \pm \sqrt{a_0^2 - 4b_1}\right).$$

Computing numerical values, with $\zeta = \exp(2i\pi/17)$, one then checks that $c_0 = 2\cos(2\pi/17)$ is given by the + sign, so that we have proved the following amazing formula:

$$2\cos(2\pi/17) = -\frac{1}{8} + \frac{1}{8}\sqrt{17} + \frac{1}{8}\sqrt{34 - 2\sqrt{17}}$$
$$+ \frac{1}{8}\sqrt{68 + 12\sqrt{17} - 2\sqrt{34 - 2\sqrt{17}} + 2\sqrt{34 - 2\sqrt{17}}\sqrt{17} - 16\sqrt{34 + 2\sqrt{17}}}.$$

5.3 Composite extensions

In this section, we study the following situation. Let K be a field; let Ω be an algebraic closure of K and let E, F be two extensions of K contained in Ω. We denote by EF the subfield of Ω generated by E and F. This is by definition the *composite extension* of E and F. Introduce also their intersection $E \cap F$, hence a diagram of fields as follows:

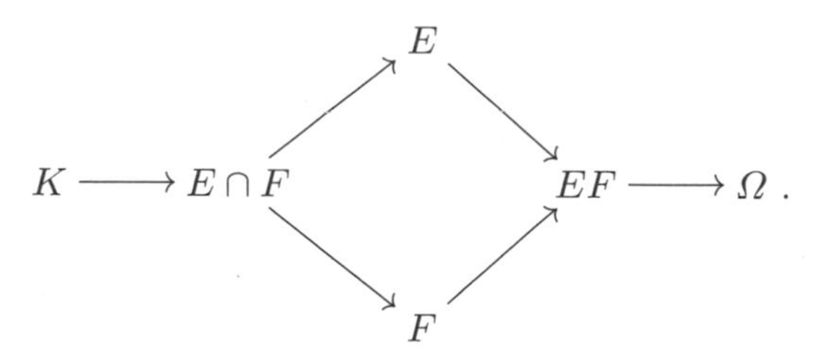

Lemma 5.3.1. *If the extension $K \subset E$ is Galois, then the extension $F \subset EF$ is Galois. If moreover the extension $K \subset F$ is Galois, the extensions $K \subset EF$ and $K \subset E \cap F$ are Galois.*

Proof. Assume that $K \subset E$ is a splitting extension of a separable polynomial $P \in K[X]$ (in other words, E is generated by the roots of P in Ω). Then $F \subset EF$ is a splitting extension of P over the field F, so is a Galois extension, by Proposition 3.2.7. If $K \subset F$ is itself a splitting extension of a separable polynomial $Q \in K[X]$, then $K \subset EF$ is a splitting extension of the polynomial PQ or, preferably, of the separable polynomial $\mathrm{l.c.m.}(P, Q)$ (see Exercise 3.2). In particular, the extension $K \subset EF$ is Galois. This shows the first two assertions of the lemma.

To prove that the extension $K \subset E \cap F$ is Galois, provided $K \subset E$ and $K \subset F$ are, it suffices to check that for any K-homomorphism $\sigma \colon E \cap F \to \Omega$, one has $\sigma(E \cap F) = E \cap F$, for then the result will follow from Proposition 3.2.7. By Theorem 3.1.6, such a morphism σ can be extended to a K-homomorphism $\tau \colon EF \to \Omega$. Since the extension $K \subset E$ is Galois, $\tau(E) = E$. Similarly, $\tau(F) = F$. Hence, $\tau(E \cap F) \subset E \cap F$. By Remark 3.2.3, $\tau(E \cap F) = E \cap F$.□

Assume now that $K \subset E$ is a Galois extension and let us show how one can identify $\mathrm{Gal}(EF/F)$ with a subgroup of $\mathrm{Gal}(E/K)$. An element $\sigma \in \mathrm{Gal}(EF/F)$ is an automorphism of EF which restricts to the identity on F. In particular, $\sigma|_K = \mathrm{id}_K$ and $\sigma \in \mathrm{Gal}(EF/K)$. Since the extension $K \subset E$ is Galois, $\sigma(E) = E$, so that σ defines an element in $\mathrm{Gal}(E/K)$ that we denote $i(\sigma)$. The map $i \colon \mathrm{Gal}(EF/F) \to \mathrm{Gal}(E/K)$ is a morphism of groups, because it is the composition of the two natural morphisms

$$\mathrm{Gal}(EF/F) \hookrightarrow \mathrm{Gal}(EF/K) \twoheadrightarrow \mathrm{Gal}(E/K).$$

Proposition 5.3.2. *The morphism i is injective; its image is* $\mathrm{Gal}(E/E \cap F)$.

Proof. If $\sigma \in \mathrm{Gal}(EF/F)$ satisfies $i(\sigma) = \mathrm{id}_E$, then σ restricts to the identity on E. One thus has $\sigma(x) = x$ for any x in F and for any x in E, so that $\sigma(x) = x$ for any x in the field generated by E and F, which is EF. This shows that $\sigma = \mathrm{id}$, hence i is injective.

Its image $i(\mathrm{Gal}(EF/F))$ is a subgroup H of $\mathrm{Gal}(E/K)$ and corresponds by Galois correspondence to the subfield E^H of E and one has $H = \mathrm{Gal}(E/E^H)$.

(Recall that E^H is the set of all $x \in E$ such that $\sigma(x) = x$ for any $\sigma \in \mathrm{Gal}(EF/F)$.) Therefore $E \cap F \subset E^H$, but conversely, if $x \in E \setminus (E \cap F)$, one has $x \in EF \setminus F$ and there is $\sigma \in \mathrm{Gal}(EF/F)$ with $\sigma(x) \neq x$, hence $x \notin E^H$. This shows that $E^H = E \cap F$; consequently $H = \mathrm{Gal}(E/E \cap F)$. □

An immediate corollary of this proposition is the following formula for the degrees of the various extensions we have been discussing.

Corollary 5.3.3. *Assume that the extension $K \subset E$ is Galois. Then,*

$$[EF : F] = [E : E \cap F].$$

In particular, $[EF : K] = [E : K]\,[F : K]$ if and only if $K = E \cap F$.

Proof. Indeed, $[EF : F]$ is the cardinality of $\mathrm{Gal}(EF/F)$. By the proposition, $\operatorname{card} i(\mathrm{Gal}(EF/F)) = \operatorname{card} \mathrm{Gal}(E/E \cap F)$, whence $[EF : F] = [E : E \cap F]$. Consequently,

$$[EF : K] = [E : E \cap F]\,[F : K] = \frac{[E : K]\,[F : K]}{[E \cap F : K]},$$

so that $[EF : K] = [E : F]\,[F : K]$ if and only if $E \cap F = K$. □

In the case where the two extensions $K \subset E$ and $K \subset F$ are Galois, we will compute the Galois group of EF over K in terms of the groups $\mathrm{Gal}(E/K)$ and $\mathrm{Gal}(F/K)$. First consider the homomorphism

$$j \colon \mathrm{Gal}(EF/K) \to \mathrm{Gal}(E/K) \times \mathrm{Gal}(F/K)$$

deduced from the two surjective morphisms $\mathrm{Gal}(EF/K) \to \mathrm{Gal}(E/K)$ and $\mathrm{Gal}(EF/K) \to \mathrm{Gal}(F/K)$. (They are well defined, for the extensions $K \subset E$ and $K \subset F$ are Galois; see Proposition 3.2.9). If $\sigma \in \mathrm{Gal}(EF/K)$ restricts to the identity on E and on F, it induces the identity on the field generated by E and F in Ω, hence on EF. It follows that $\sigma = \mathrm{id}$ and j is injective.

First assume that $K = E \cap F$. By Corollary 5.3.3, one has $[EF : K] = [E : K][F : K]$. Since j is injective, it must be surjective.

In the general case, we have shown in Lemma 5.3.1 that the extension $K \subset E \cap F$ is Galois. Let us see what happens if we compose j with the surjective homomorphisms

$$\pi_1 \colon \mathrm{Gal}(E/K) \times \mathrm{Gal}(F/K) \to \mathrm{Gal}(E/K) \to \mathrm{Gal}(E \cap F/K)$$

and

$$\pi_2 \colon \mathrm{Gal}(E/K) \times \mathrm{Gal}(F/K) \to \mathrm{Gal}(F/K) \to \mathrm{Gal}(E \cap F/K).$$

By construction, $\pi_1 \circ j$ and $\pi_2 \circ j$ are both equal to the natural morphism

$$\mathrm{Gal}(EF/K) \to \mathrm{Gal}(E \cap F/K)$$

corresponding to the Galois subextension $K \subset E \cap F$ of $K \subset EF$. Therefore, the image of j is contained in the subgroup G of $\mathrm{Gal}(E/K) \times \mathrm{Gal}(F/K)$ consisting of all (σ_1, σ_2) such that $\pi_1(\sigma_1) = \pi_2(\sigma_2)$.

If we show that $\operatorname{card} G = \operatorname{card} \mathrm{Gal}(EF/K)$, it will follow that j is an isomorphism from $\mathrm{Gal}(EF/K)$ onto G. If Δ denotes the diagonal subgroup in $\mathrm{Gal}(E \cap F/K) \times \mathrm{Gal}(E \cap F/K)$ consisting of all (σ, σ), with $\sigma \in \mathrm{Gal}(E \cap F/K)$, we see that G is the preimage of Δ by the surjective morphism

$$(\pi_1, \pi_2)\colon \mathrm{Gal}(E/K) \times \mathrm{Gal}(F/K) \to \mathrm{Gal}(E \cap F/K) \times \mathrm{Gal}(E \cap F/K).$$

Therefore,

$$\begin{aligned}
\operatorname{card} G &= \operatorname{card} \mathrm{Gal}(E \cap F/K) \times \operatorname{card} \mathrm{Ker}(\pi_1, \pi_2) \\
&= [E \cap F : K] \times [E : E \cap F] \times [F : E \cap F] \\
&= [F : K] \times [E : E \cap F] \\
&= [F : K] \times [EF : F] \\
&= [EF : K].
\end{aligned}$$

We have thus proved the following theorem.

Theorem 5.3.4. *Let us consider a composite extension $K \subset EF$, where $K \subset E$ and $K \subset F$ are two Galois extensions contained in an algebraic closure of K. The extension $K \subset EF$ is Galois and its Galois group is isomorphic to the subgroup of* $\mathrm{Gal}(E/K) \times \mathrm{Gal}(F/K)$ *consisting of all couples (σ, τ) such that σ and τ have the same image in* $\mathrm{Gal}(E \cap F/K)$.

In particular, if $E \cap F = K$, $\mathrm{Gal}(EF/K)$ *can be identified with the direct product* $\mathrm{Gal}(E/K) \times \mathrm{Gal}(F/K)$.

5.4 Cyclic extensions

By definition, a *cyclic extension* is a Galois extension whose Galois group is cyclic, hence isomorphic to $\mathbf{Z}/n\mathbf{Z}$, where n denotes the degree of the extension.

If K is a field, let us denote by $\mu_n(K)$, or μ_n in short, the (cyclic) group of nth roots of unity in K. In this section, we will often assume that μ_n has order n. In this case, it is generated by a primitive nth root of unity. If the characteristic of the field K is equal to a prime number p, this implies that n is not a multiple of p.

This section is devoted to the determination of the field extensions of K which are Galois with Galois group $\mathbf{Z}/n\mathbf{Z}$.

Let us begin by an example, which in fact is *the* example.

Theorem 5.4.1. *Let K be a field and let n be any integer with $n \geqslant 2$. We assume that* $\operatorname{card} \mu_n(K) = n$.

Let $a \in K^*$ and let $K \subset L$ be a splitting extension of the polynomial $P = X^n - a$; denote by x a root of P in L.

The extension $K \subset L$ is Galois. The map $i\colon \sigma \mapsto i(\sigma) = \sigma(x)/x$ defines an injective group morphism from $\operatorname{Gal}(L/K)$ *to $\mu_n(K)$. Let d be the smallest positive integer such that $x^d \in K$. Then d divides n, and the image of the morphism i is equal to $\mu_d(K)$.*

In particular, the following are equivalent:

a) *for any integer $m > 1$ dividing n, a is not an mth power in K;*
b) *the polynomial $X^n - a$ is irreducible in $K[X]$;*
c) $\operatorname{Gal}(L/K) \simeq \mathbf{Z}/n\mathbf{Z}$.

Proof. In $L[X]$, the polynomial $P = X^n - a$ can be factored as

$$P = X^n - a = \prod_{\zeta \in \mu_n} (X - \zeta x).$$

Since $\operatorname{card} \mu_n(K) = n$, the characteristic of K does not divide n and any root ζx of P in L is simple, for $P'(\zeta x) = n(\zeta x)^{n-1} = na/(\zeta x) \neq 0$. In other words, the polynomial P is separable and the extension $K \subset L$ is Galois.

Any K-automorphism σ of L is determined by the image $\sigma(x)$ of x, which is a root of $X^n - a$. Then $\sigma(x)/x$ is an nth root of unity. This defines a map $i\colon \operatorname{Gal}(L/K) \to \mu_n$, such that $i(\sigma) = \sigma(x)/x$.

Observe that i is a group homomorphism; if $\sigma(x) = ux$ and $\tau(x) = vx$ for u, $v \in \mu_n$, then

$$(\sigma \circ \tau)(x) = \sigma(vx) = v\sigma(x) = uvx,$$

since $v \in K$; hence $i(\sigma \circ \tau) = i(\sigma)i(\tau)$. The image of i in μ_n is a subgroup, necessarily of the form μ_d for some integer d dividing n. One has $[L : K] = \operatorname{card} \operatorname{Gal}(L/K) = d$, and d is the degree of the minimal polynomial of x over K, for $L = K[x]$. Notice also that $\operatorname{Gal}(L/K) \simeq \mu_d(K) \simeq \mathbf{Z}/d\mathbf{Z}$.

Let m be any integer. One has $x^m \in K$ if and only if $\sigma(x^m) = x^m$ for any $\sigma \in \operatorname{Gal}(L/K)$. Since $\sigma(x) = i(\sigma)x$, this holds if and only if $i(\sigma)^m = 1$ for any $\sigma \in \operatorname{Gal}(L/K)$, hence if and only if $\zeta^m = 1$ for any ζ in the image of $\operatorname{Gal}(L/K)$ by i. Since $i(\operatorname{Gal}(L/K)) = \mu_d(K)$, one has $x^m \in K$ if and only if d divides m. It follows in particular that $a = x^n = (x^d)^{n/d}$ is an (n/d)th power in K. If one assumes that a is not an mth power in K for any integer $m > 1$ dividing n, then $d = n$ and $P = X^n - a$ is irreducible in $K[X]$. Conversely, assuming that $a = b^e$ for some $b \in K$ and some integer $e > 1$ dividing n, the equality

$$P = X^n - a = X^{me} - b^e = (X^m - b)(X^{m(e-1)} + X^{m(e-2)}b + \cdots + b^{e-1})$$

shows that P is not irreducible in $K[X]$. □

Conversely, let $K \to L$ be any finite Galois extension, with Galois group $\mathbf{Z}/n\mathbf{Z}$. Let σ be any generator of $\operatorname{Gal}(L/K)$. The preceding proof suggests that we seek for an element $x \in L$ such that $L = K[x]$ and such that $\sigma(x)/x$ is an nth root of unity. Let $\zeta \in \mu_n$ be any primitive nth root of unity and let us define, for $t \in L$, the *Lagrange's resolvent* of t, by the formula

$$x = t + \zeta^{-1}\sigma(t) + \cdots + \zeta^{1-n}\sigma^{n-1}(t).$$

It is proved in Exercise 3.12 that the elements of $\operatorname{Gal}(L/K)$ are linearly independent over K. Consequently, one may find $t \in L$ such that $x \neq 0$. Then

$$\sigma(x) = \sigma(t) + \zeta^{-1}\sigma^2(t) + \cdots + \zeta^{1-n}\sigma^n(t) = \zeta x,$$

since $\sigma^n = \mathrm{id}$ and $\zeta^n = 1$. By induction, for any $k \in \{0, 1, \ldots, n-1\}$, one has

$$\sigma^k(x) = \zeta^k x.$$

It follows that for any integer k, $\sigma^k(x^n) = x^n$. Since $\operatorname{Gal}(L/K) = \{\mathrm{id}, \sigma, \sigma^2, \ldots, \sigma^{n-1}\}$, $a = x^n$ belongs to K.

Let

$$P = X^n - a = \prod_{k=0}^{n-1}(X - \zeta^k x) = \prod_{k=0}^{n-1}(X - \sigma^k(x)).$$

It is a separable polynomial in $K[X]$, split in L. Consequently, the extension $K \subset K(x)$ is a splitting extension of the polynomial P. Since $\operatorname{Gal}(L/K)$ acts transitively on its roots, this polynomial is irreducible. This implies $[K(x) : K] = n$, and since one has $[L : K] = n$, it follows necessarily that $L = K(x)$, hence x is a primitive element of the extension $K \subset L$, with minimal polynomial $X^n - a$. (See also Exercise 3.8.)

We finally have proved the following theorem.

Theorem 5.4.2. *Let K be a field and let n be any integer $\geqslant 2$ such that $\operatorname{card}\mu_n(K) = n$.*

If $K \subset L$ is a Galois extension whose Galois group is isomorphic to $\mathbf{Z}/n\mathbf{Z}$, there exists $a \in K$ such that L is a splitting extension of the irreducible polynomial $X^n - a \in K[X]$.

5.5 Equations with degrees up to 4

We are now going to analyse equations of degree $\leqslant 4$ in light of Galois theory. What will allow us to explicitly solve such equations is that in each of the three groups $\mathfrak{S}_2$, $\mathfrak{S}_3$ and $\mathfrak{S}_4$, there is a normal series such that the successive quotients are cyclic (with order 2 or 3), i.e. , these groups are solvable. By Corollary 4.6.8, this does not happen in $\mathfrak{S}_n$ for $n > 4$. Recall that the symbol $\lhd$ means that the group on the left is normal in the next, and the number above it indicates the order of the quotient. Also recall that one denotes by V_4 the *Klein four-group* in $\mathfrak{A}_4$, the four elements of which are the permutations

$$\text{id}, \quad (1,2)(3,4), \quad (1,3)(2,4), \quad (1,4)(2,3)$$

on the set $\{1,2,3,4\}$ (these are the products of two transpositions with disjoint supports, plus the identity). This group is isomorphic to $(\mathbf{Z}/2\mathbf{Z})^2$. Now, one has the following normal series:

$$\begin{aligned} \{1\} = \mathfrak{A}_2 &\overset{2}{\lhd} \mathfrak{S}_2 = \mathbf{Z}/2\mathbf{Z} \\ \{1\} &\overset{3}{\lhd} \langle(1,2,3)\rangle \overset{2}{\lhd} \mathfrak{A}_3 \overset{2}{\lhd} \mathfrak{S}_3 \\ \{1\} \overset{2}{\lhd} \{1,(1,2)(3,4)\} &\overset{2}{\lhd} V_4 \overset{3}{\lhd} \mathfrak{A}_4 \overset{2}{\lhd} \mathfrak{S}_4 \ . \end{aligned}$$

Felix Klein (1849–1925)

In this section, we consider only fields whose characteristic is neither 2 *nor* 3.

Let K be such a field and let P be a monic polynomial in $K[X]$ with degree $n \leqslant 4$. Let $K \subset L$ be the splitting extension of P contained in some fixed algebraic closure Ω of K. (All extensions in this section will be assumed to live in Ω.) Denote by $x_1, x_2, \ldots, x_n$ the roots of P in L and let $G = \text{Gal}(L/K)$. This is naturally a subgroup of $\mathfrak{S}_n$.

The intersections with G of the above-written subgroups of $\mathfrak{S}_n$ define a normal series in G, and the successive quotients are cyclic groups with order $\leqslant 3$ (they may be trivial). Such a series corresponds to a chain of Galois extensions. We already explained in Chapter 3, Prop. 3.4.2, how the subgroup $\mathfrak{A}_n \subset \mathfrak{S}_n$ corresponds to the extension generated by a square root of the discriminant of P.

Let us first consider degree 2. Then $P = X^2 + aX + b$ for $a, b \in K$ and the discriminant of P is equal to $\Delta = a^2 - 4b$. If Δ is a square in K, the roots of P belong to K and $G = \{1\}$. Otherwise, $L = K(\sqrt{\Delta})$ has degree 2 over K. We can order the roots so that $x_1 - x_2 = \sqrt{\Delta}$. Together with the relation $x_1 + x_2 = a$, this determines $x_1 = (a + \sqrt{\Delta})/2$ and $x_2 = (a - \sqrt{\Delta})/2$.

Assume now that P is a separable polynomial with degree 3 in $K[X]$:

$$P = X^3 + a_1 X^2 + a_2 X + a_3.$$

The change of variables $Y = X + a_1/3$ allows us to assume that the sum of its roots is equal to 0 or, in other words, that P is of the form $P = X^3 + pX + q$. Its discriminant is then equal to $D = -4p^3 - 27q^2$ (Example 3.4.1). Let us consider the extensions

$$K \overset{2}{\subset} K(\sqrt{\Delta}) \overset{3}{\subset} L,$$

where each extension is either trivial, or Galois with Galois group the cyclic group of cardinality indicated above the inclusion sign. If the polynomial P is irreducible, we already can deduce from this the Galois group of L over K. Indeed, the degree of the extension $K \subset L$ is a multiple of 3 and $\mathrm{Gal}(L/K)$ is $\mathfrak{S}_3$ when Δ is not a square, and is $\mathfrak{A}_3$ if Δ is a square in K.

To give explicit formulas for the roots of P, we first have to adjoin $\sqrt{\Delta}$. The remaining extension $K(\sqrt{\Delta}) \overset{3}{\subset} L$ is either trivial if the field $K(\sqrt{\Delta})$ contains the three roots of P, or cyclic with Galois group $\mathbf{Z}/3\mathbf{Z}$.

Proceeding as in the case of extensions with a cyclic Galois group (Section 5.4), let us first add to $K(\sqrt{\Delta})$ the cubic roots of unity ρ and ρ^2. These are the roots of the polynomial $X^2 + X + 1$. Recall that we may assume

$$\rho = -\frac{1}{2} + \frac{1}{2}\sqrt{-3}, \quad \rho^2 = -\frac{1}{2} - \frac{1}{2}\sqrt{-3}$$

where $\sqrt{-3}$ denotes a square root of -3 in $K(\sqrt{\Delta}, \rho)$. In particular, $\rho - \rho^2 = \sqrt{-3}$. Set $K' = K(\rho)$ and $L' = L(\rho)$.

The resulting extension $K'(\sqrt{\Delta}) \subset L'$ is either trivial or cyclic with order 3. Corresponding to the circular permutation $(1, 2, 3)$, there are two *Lagrange's resolvents* that one can introduce:

$$\alpha = x_1 + \rho x_2 + \rho^2 x_3 \quad \text{and} \quad \beta = x_1 + \rho^2 x_2 + \rho x_3.$$

Let us now compute α^3 and β^3:

$$\alpha^3 = x_1^3 + x_2^3 + x_3^3 + 6x_1x_2x_3 + 3\rho(x_1^2x_2 + x_2^2x_3 + x_3^2x_1) + 3\rho^2(x_1x_2^2 + x_2x_3^2 + x_3x_1^2)$$

and β^3 is given by the formula obtained by switching ρ and ρ^2. The first term in these expressions is a symmetric function of the roots, hence can be expressed with p and q. Explicitly:

$$\begin{aligned} x_1^3 + x_2^3 + x_3^3 + 6x_1x_2x_3 &= (x_1 + x_2 + x_3)^3 - 3(x_1^2x_2 + \dots) \\ &= -3\big(x_1x_2(x_1 + x_2) + \dots\big) \\ &= -3x_1x_2(-x_3) - \dots \\ &= 9x_1x_2x_3 = -9q. \end{aligned}$$

The two other terms are not symmetric functions, and we cannot hope for them to be, otherwise α^3 and β^3 would always belong to K'. However, we *know* that they belong to $K'(\sqrt{\Delta})$ and the aim of the game is to find a formula for them! Since Δ has two square roots, we have to choose one of them and we set

$$\sqrt{\Delta} = (x_1-x_2)(x_1-x_3)(x_2-x_3) = (x_1^2x_2+x_2^2x_3+x_3^2x_1)-(x_1x_2^2+x_2x_3^2+x_3x_1^2).$$

Defining

$$A = x_1^2x_2 + x_2^2x_3 + x_3^2x_1 \quad \text{and} \quad B = x_1x_2^2 + x_2x_3^2 + x_3x_1^2,$$

we find the relations

$$A + B = 3q \quad \text{and} \quad A - B = \sqrt{\Delta},$$

hence

$$A = \frac{3}{2}q + \frac{1}{2}\sqrt{\Delta} \quad \text{and} \quad B = \frac{3}{2}q - \frac{1}{2}\sqrt{\Delta}.$$

Let us write down these expressions in the formulae for α^3 and β^3:

$$\begin{aligned}
\alpha^3 &= -9q + 3\rho A + 3\rho^2 B \\
&= -9q + \frac{3}{2}q(3\rho + 3\rho^2) + \frac{1}{2}\sqrt{\Delta}(3\rho - 3\rho^2) \\
&= -\frac{27}{2}q + \frac{3}{2}\sqrt{-3\sqrt{\Delta}}
\end{aligned}$$

and

$$\beta^3 = -\frac{27}{2}q - \frac{3}{2}\sqrt{-3}\sqrt{\Delta}.$$

Moreover, since $\sigma(\alpha) = \rho^{-1}\alpha$ and $\sigma(\beta) = \rho^{-2}\beta$, one has $\sigma(\alpha\beta) = \alpha\beta$ and $\alpha\beta \in K'$. Actually,

$$\begin{aligned}
\alpha\beta &= (x_1 + \rho x_2 + \rho^2 x_3)(x_1 + \rho^2 x_2 + \rho x_3) \\
&= x_1^2 + x_2^2 + x_3^2 + (\rho + \rho^2)(x_1x_2 + x_2x_3 + x_3x_1) \\
&= (x_1 + x_2 + x_3)^2 + (\rho + \rho^2 - 2)(x_1x_2 + x_2x_3 + x_3x_1) \\
&= -3p.
\end{aligned}$$

To derive explicit formulae for x_1, x_2 and x_3, it remains to note that one has a Cramer system with three linear equations in three unknowns:

$$\begin{cases} x_1 + x_2 + x_3 = 0 \\ x_1 + \rho x_2 + \rho^2 x_3 = \alpha \\ x_1 + \rho^2 x_2 + \rho x_3 = \beta. \end{cases}$$

Jerome Cardan (1501–1576)

Therefore,

$$\begin{cases} x_1 = \frac{1}{3}\alpha + \frac{1}{3}\beta \\ x_2 = \frac{1}{3}\rho^2\alpha + \frac{1}{3}\rho\beta \\ x_3 = \frac{1}{3}\rho\alpha + \frac{1}{3}\rho^2\beta. \end{cases}$$

These are "*Cardan's formulae.*" (Concerning history, Jerome Cardan had bought them to Tartaglia under the promise of not publishing them, a promise which was broken when Cardan published his *Ars magna sive de regulis algebraicis liber unus* in 1545. Before that, Scipione del Ferro, an Italian like Cardan, had discovered how to solve equations of degree 3 but only at the moment of his death did he explain his method, and only for a particular case!)

In practice, if one wants to solve a cubic equation, this can all be ignored and one needs to remember only the following procedure: write one of the roots $x = u + v$ with $uv = -p/3$, then expand

$$0 = (u+v)^3 + p(u+v) + q = u^3 + v^3 + 3uv(u+v) + p(u+v) + q = u^3 + v^3 + q,$$

so that u^3 and v^3 are solutions of the quadratic equation

$$X^2 + qX - \frac{p^3}{27} = 0.$$

Therefore, the value of u^3 can be deduced from one of the square roots of the discriminant $q^2 + \frac{4}{27}p^3 = -\Delta/27$, then the value of u through a cubic root, and finally the value for $x = u - p/3u$. (This works well for $p \neq 0$, but the case $p = 0$ is easy.)

You might also notice that when x_1, x_2 and x_3 are real numbers, Δ is a positive real number, hence Cardan's formulae use complex numbers. This is the so-called *casus irreductibilis*, and there is no way to avoid it (see Exercise 7.2).

Let us finally explain how to solve equations of degree 4. Let

$$P = X^4 + pX^2 + qX + r$$

be a monic polynomial of degree 4, where the coefficient of X^3 is assumed to be 0 up to a linear change of variables. Let us recall the sequence of normal subgroups in $\mathfrak{S}_4$:

$$\{1\} \overset{2}{\triangleleft} \{1, (1,2)(3,4)\} \overset{2}{\triangleleft} V_4 \overset{3}{\triangleleft} \mathfrak{A}_4 \overset{2}{\triangleleft} \mathfrak{S}_4,$$

hence a chain of Galois extensions

$$K \overset{2}{\subset} K(\sqrt{\Delta}) \overset{3}{\subset} K_1 \overset{2}{\subset} K_2 \overset{2}{\subset} L.$$

(The numbers above the inclusion symbol mean that the extension is either trivial, or cyclic with corresponding degree.) We now can use a similar approach to the one we gave for degree 3.

Let us immediately introduce a resolvent polynomial corresponding to the extension $K \subset K_1$. The polynomial $R_1 = (X_1 + X_2)(X_3 + X_4)$ is invariant under the permutations of V_4, and its orbit under the symmetric group $\mathfrak{S}_4$ consists of the three polynomials

$$R_1, \quad R_2 = (X_1 + X_3)(X_2 + X_4) \quad \text{and} \quad R_3 = (X_1 + X_4)(X_2 + X_3).$$

It follows first that $\theta_1 = (x_1 + x_2)(x_3 + x_4)$, $\theta_2 = (x_1 + x_3)(x_2 + x_4)$, and $\theta_3 = (x_1 + x_4)(x_2 + x_3)$ belong to $K_1 = L^{V_4}$, and second that the degree 3 polynomial

$$Q(X) = (X - \theta_1)(X - \theta_2)(X - \theta_3)$$

has its coefficients in K. This polynomial is usually called the Lagrange's resolvent polynomial of the quartic equation. If P is separable, which we assume, then θ_1, θ_2 and θ_3 are distinct. In fact, one has

$$\theta_1 - \theta_2 = (x_4 - x_1)(x_2 - x_3),$$

and similar formulae for $\theta_2 - \theta_3$ and $\theta_1 - \theta_2$. This shows moreover that the discriminant of Q is equal to that of P.

Exercise 5.5.1. Show that $Q(X) = X^3 - 2pX^2 + (p^2 - 4r)X + q^2$.

Assume now that we have determined θ_1, θ_2 and θ_3, e.g. , using Cardan's formulae. By the relations $(x_1+x_2)(x_3+x_4) = \theta_1$ and $(x_1+x_2)+(x_3+x_4) = 0$, we see that $x_1 + x_2$ is a square root of $-\theta_1$, say $\sqrt{-\theta_1}$. Similarly, $x_1 + x_3$ and $x_1 + x_4$ are square roots of $-\theta_2$ and $-\theta_3$ respectively. Pay attention to the fact that these three square roots cannot be taken arbitrarily: the degree of the extension $K_1 \subset L$ divides 4 and three "independent" square roots would make the degree a multiple of 8. Actually, one has

$$\begin{aligned}\sqrt{-\theta_1}\sqrt{-\theta_2}\sqrt{-\theta_3} &= (x_1+x_2)(x_1+x_3)(x_1+x_4)\\ &= x_1^3 + x_1^2(x_2+x_3+x_4) + x_1x_2x_3 + x_1x_2x_4 + x_1x_3x_4 + x_2x_3x_4 = -q.\end{aligned}$$

If $q = 0$, the quartic equation is "biquadratic" and can be solved easily. Otherwise, if $q \neq 0$, the θ_i are nonzero and this formula computes $\sqrt{-\theta_3} = -q/\sqrt{-\theta_1}\sqrt{-\theta_2}$. Then,

$$2x_1 = 3x_1 + x_2 + x_3 + x_4 = \sqrt{-\theta_1} + \sqrt{-\theta_2} + \sqrt{-\theta_3}$$

and analogous formulae for x_2, x_3 and x_4.

Assuming that P is irreducible in $K[X]$, let us determine the various possible Galois groups.

First observe that the extension $K(\sqrt{\Delta}) \subset K_1$ has degree either 1 or 3, for it is Galois and its Galois group is a subquotient of $\mathfrak{A}_4/V_4 \simeq \mathbf{Z}/3\mathbf{Z}$. This shows that the polynomial Q cannot have an irreducible factor of degree 2 over $K(\sqrt{\Delta})$. Therefore, it is either split or irreducible over $K(\sqrt{\Delta})$. In this last case, the degree $[L:K]$ is divisible by 3 and by Cauchy's lemma (Proposition 4.2.3, but you might want to prove it by hand here), $\mathrm{Gal}(L/K)$ contains an element of order 3, hence a subgroup of order 3. But there are precisely four such subgroups in $\mathfrak{S}_4$, denoted $H_1, \dots, H_4$, where H_i is generated by any 3-cycle which fixes i. (For example, H_1 is generated by the 3-cycle $(2,3,4)$.) If $g \in \mathfrak{S}_4$ maps i to j, then $gH_jg^{-1} = H_i$; since $\mathrm{Gal}(L/K)$ acts transitively on $\{1,2,3,4\}$, as soon as $\mathrm{Gal}(L/K)$ contains one of the H_i, it contains the other three, hence all 3-cycles, hence all of $\mathfrak{A}_4$. We just proved that if Q is irreducible over $K(\sqrt{\Delta})$, then $\mathrm{Gal}(L/K)$ contains $\mathfrak{A}_4$.

If moreover Δ is a square in K, one has $\mathrm{Gal}(L/K) \subset \mathfrak{A}_4$, whence the equality. If Δ is not a square in K, one has $\mathrm{Gal}(L/K) = \mathfrak{S}_4$.

Let us now assume that the resolvent polynomial Q is split in $K(\sqrt{\Delta})$, i.e. , assume that $K_1 = K(\sqrt{\Delta})$. Since P is irreducible, $[L:K]$ is a multiple of 4. Moreover, $[L:K]$ divides 8, hence one has $[L:K] = 4$ or 8.

If Δ is a square in K, one then has $K_1 = K$ and $\mathrm{Gal}(L/K) \subset V_4$. Since no proper subgroup of V_4 acts transitively on $\{1,2,3,4\}$, one necessarily has $\mathrm{Gal}(L/K) = V_4$ in this case.

If Δ is not a square in K, one has $[K_1 : K] = 2$. Therefore, $[L:K] = 4$ if L is generated by one of the square roots of the $-\theta_j$, and $[L:K] = 8$ otherwise. In the first case, $\mathrm{Gal}(L/K)$ is a transitive subgroup of order 4 in $\mathfrak{S}_4$, not contained in $\mathfrak{A}_4$, which leaves only the cyclic group generated by a circular permutation. In the other case, $\mathrm{Gal}(L/K)$ has 8 elements and is isomorphic to the dihedral group D_4. (Remark: it is one of the 2-Sylow subgroups of $\mathfrak{S}_4$, generated by a 4-cycle (a,b,c,d) and the transposition (a,c).)

5.6 Solving equations by radicals

In this section, I explain the relationship discovered by Galois between the possibility of solving a polynomial equation with radicals and the solvability of its Galois group. This relationship simultaneously generalizes the following results:

– Theorem 5.1.1 concerning constructibility with ruler and compass (on one hand, a group with cardinality a power of 2 is solvable, see Exercise 4.11; on the other hand, constructible numbers are contained in an extension obtained by successively adding square roots);

– the explicit solution of equations with degree 2, 3 or 4 which I explained in the previous section (as I said there, the groups $\mathfrak{S}_2$, $\mathfrak{S}_3$ and $\mathfrak{S}_4$, and their subgroups, are solvable);

– Abel's theorem (see Corollary 5.6.5 below) that the general equation of degree $n \geqslant 5$ is not solvable by radicals.

To simplify, we will only consider in this section fields of characteristic zero.

Definition 5.6.1. *Let E be a field with characteristic zero, and let $E \subset F$ be a finite extension.*

We will say that the extension $E \subset F$ is elementary radical *with exponent n, if there is some $x \in F$ such that $F = E[x]$ and $x^n \in E$.*

We will say that the extension $E \subset F$ is radical *if there are subfields $E = E_0 \subset E_1 \subset \cdots \subset E_n = F$ such that the extension $E_{i-1} \subset E_i$ is elementary radical for any $i \in \{1, \ldots, n\}$.*

Finally, we will say that the extension $E \subset F$ is solvable by radicals, *or simply* solvable, *if there is a finite extension $F \subset F'$ such that the extension $E \subset F'$ is a radical extension.*

Proposition 5.6.2. a) *Let $E \subset F$ be a finite extension and K be any field such that $E \subset K \subset F$. If the extension $E \subset F$ is radical, then the extension $K \subset F$ is itself radical. If the extension $E \subset F$ is solvable, then both extensions $E \subset K$ and $K \subset F$ are solvable.*

b) *Let $E \subset F_1$ and $E \subset F_2$ be two finite isomorphic extensions. If $E \subset F_1$ is a radical extension (*resp. *a solvable extension), then so is $E \subset F_2$.*

c) *Let Ω be a field containing E and let $E \subset F$ and $E \subset F'$ be two radical (*resp. *solvable) extensions contained in Ω. Then the composite extension $E \subset FF'$ is radical (*resp. *solvable).*

d) *Let $E \subset F$ a finite radical (*resp. *solvable) extension. Then its Galois closure (in any algebraic closure), $E \subset F^{\mathrm{g}}$, is again radical (*resp. *solvable).*

Proof. *a*) is obvious from the Definition.

b) Assume the extension $E \subset F_1$ to be radical. Let $E = E_0 \subset E_1 \subset \cdots \subset E_n = F_1$ be a chain of subfields such that the extension $E_{i-1} \subset E_i$ is elementary radical for any integer i. Let $\sigma\colon F_1 \to F_2$ be any E-isomorphism. For any i, the extension $\sigma(E_{i-1}) \subset \sigma(E_i)$ is elementary radical, for if $E_i = E_{i-1}(x_i)$, with $x_i^{n_i} \in E_{i-1}$, one has $\sigma(E_i) = \sigma(E_{i-1})(\sigma(x_i))$ and $\sigma(x_i)^{n_i} \in \sigma(E_{i-1})$. This shows that the extension $\sigma(E) \subset \sigma(F_1)$ is radical.

Now assume that the extension $E \subset F_1$ is solvable and let F_1' be some extension of F_1 such that the extension $E \subset F_1'$ is radical. Fix an algebraic closure Ω of F_2; by Theorem 3.1.6, there is a field homomorphism $\sigma'\colon F_1' \to \Omega$ such that $\sigma'|_{F_1} = \sigma$. Consequently, the extension $E \subset \sigma'(F_1')$ is radical, and the extension $E \subset F_2$ is solvable.

c) Let $E = E_0 \subset E_1 \subset \cdots \subset E_n = F$ and $E = E_0' \subset E_1' \subset \cdots \subset E_{n'}' = F'$ be two chains of fields, the extensions $E_{i-1} \subset E_i$ and $E_{i-1}' \subset E_i'$ being elementary radical. If y_i is an element in E_i' such that $E_i' = E_{i-1}'(y_i)$, a power of which belongs to E_{i-1}', then the extension $FE_{i-1}' \subset FE_i'$ is elementary radical, for $FE_i' = FE_{i-1}'(y_i)$. The chain of elementary radical extensions

$$E = E_0 \subset E_1 \subset \cdots \subset E_n = F \subset FE_1' \subset FE_2' \subset \cdots \subset FE_n' = FF'$$

shows that the extension $E \subset FF'$ is radical.

Assume that the two extensions $E \subset F$ and $E \subset F'$ are solvable, and let $F \subset L$ and $F' \subset L'$ be extensions such that $E \subset L$ and $E \subset L'$ are radical. By assumption, the fields F and F' are contained in Ω, which we can assume to be algebraically closed. (Otherwise, replace Ω by any algebraic closure.) Then there is an F-homomorphism $\sigma\colon L \to \Omega$ and a F'-homomorphism $\sigma'\colon L' \to \Omega$. By *b*), the extensions $E \subset \sigma(L)$ and $E \subset \sigma'(L')$ are radical, and so is the extension $E \subset \sigma(L)\sigma'(L')$. Since $E \subset FF' \subset \sigma(L)\sigma'(L')$, the extension $E \subset FF'$ is solvable.

d) Let Ω be an algebraic closure of F. The Galois closure of an extension $E \subset F$ is the subfield of Ω generated by all $\sigma(F)$, with σ running along the set of all E-homomorphisms from F to Ω. By b), each extension $E \subset \sigma(F)$ is radical (*resp.* solvable). An obvious induction using c) now shows that the extension $E \subset \prod_\sigma \sigma(F)$ is radical (*resp.* solvable). □

Theorem 5.6.3. *Let E be a field of characteristic zero. A Galois extension $E \subset F$ is solvable if and only if its Galois group* $\mathrm{Gal}(F/E)$ *is solvable.*

Before proving this very important theorem, it might be worth recalling the Galois theory of elementary radical extensions given by Theorem 5.4.1 and its converse, Theorem 5.4.2, showing that that Galois extensions with Galois group $\mathbf{Z}/n\mathbf{Z}$ are elementary radical, since we assumed that $\operatorname{card} \mu_n(E) = n$.

Proposition 5.6.4. *Let E be a field such that* $\operatorname{card} \mu_n(E) = n$.

Any elementary radical extension $E \subset F$ of exponent n, is Galois and $\mathrm{Gal}(F/E)$ *can be identified to a subgroup of* $\mathbf{Z}/n\mathbf{Z}$. *(It follows that there is an integer d dividing n such that* $\mathrm{Gal}(F/E) \simeq \mathbf{Z}/d\mathbf{Z}$.*)*

Conversely, any Galois extension $E \subset F$ with Galois group $\mathbf{Z}/n\mathbf{Z}$ *is elementary radical, of exponent n.*

The *Proof of Theorem 5.6.3* involves four steps.

a) Let the extension $E \subset F$ be radical, Galois, and assume that E contains a root of unity of order $[F : E]$. Then $\mathrm{Gal}(F/E)$ *is a solvable group.*

Let us show this by induction on the degree $[F : E]$. Let $E \subset E_1 \subset \cdots \subset F$ be a chain of (nontrivial) elementary radical extensions. Set $G = \mathrm{Gal}(F/E)$ and $H = \mathrm{Gal}(F/E_1)$. The extension $E_1 \subset F$ is radical and Galois. Since $[F : E_1]$ and $[E_1 : E]$ both divide $[F : E]$, E contains a primitive root of unity of both orders. By induction, the group H is solvable; by the preceding proposition, the extension $E \subset E_1$ is Galois and its Galois group is cyclic. Consequently, H is a normal subgroup of G and $G/H \simeq \mathrm{Gal}(E/E_1)$ is a cyclic finite group. It now follows from Proposition 4.5.2, c), that G is a solvable group.

b) Let $E \subset F$ be a solvable Galois extension, then $\mathrm{Gal}(F/E)$ *is a solvable group.*

Let $F \subset F_1$ be a finite extension such that the extension $E \subset F_1$ is a radical extension. Let Ω be an algebraic closure of K containing F_1 and let L denote the Galois closure of the extension $E \subset F_1$ in Ω. The extension $E \subset L$ is radical and Galois. Denote also by K the extension of E generated in Ω by a primitive root of unity of order $[L : E]$.

By Proposition 5.6.2, c), the extension $K \subset KL$ is radical and Galois. Since its degree $[KL : K]$ divides $[L : E]$, a) implies that $\mathrm{Gal}(KL/K)$ is a solvable group. On the other hand, the extension $E \subset K$ is Galois, and its Galois group is a subgroup of $(\mathbf{Z}/N\mathbf{Z})^*$, where $N = [L : E]$ (see Section 5.2). Therefore, $\mathrm{Gal}(KL/K)$ is a normal subgroup of $\mathrm{Gal}(KL/E)$ and the quotient $\mathrm{Gal}(KL/E)/\mathrm{Gal}(KL/K)$ is abelian, because it is isomorphic to $\mathrm{Gal}(K/E)$. Since $\mathrm{Gal}(KL/K)$ is solvable, it follows from Prop. 4.5.2, c), that the group $\mathrm{Gal}(KL/E)$ is solvable. Since $E \subset F$ is a Galois extension with $F \subset KL$, $\mathrm{Gal}(F/E)$ is a quotient of $\mathrm{Gal}(KL/E)$. This shows that $\mathrm{Gal}(F/E)$ is a solvable group.

c) If $\mathrm{Gal}(F/E)$ *is a solvable group, and if E contains a primitive root of unity of order $[F : E]$. Then the extension $E \subset F$ is radical.*

Let us show this by induction on $[F : E]$. The group $G = \mathrm{Gal}(F/E)$ is solvable. By Proposition 4.5.3, G has a normal subgroup H, such that G/H is cyclic. Consequently, there exists an integer $d > 1$ dividing $[F : E]$ such that G/H is isomorphic to $\mathbf{Z}/d\mathbf{Z}$. Therefore, the field extension $E \subset F^H$ is

Galois, and its Galois group is $\mathbf{Z}/d\mathbf{Z}$; by Proposition 5.6.4, this extension is elementary radical. (Observe that E contains a primitive dth root of unity.) The extension $F^H \subset F$ is Galois and its Galois group is equal to H, so is solvable (Proposition 4.5.2, a). Since $[F : F^H]$ divides $[F : E]$, F^H contains a primitive root of unity of order $[F : F^H]$. By induction, the extension $F^H \subset F$ is a radical extension. This shows that the extension $E \subset F$ is radical.

d) If $\mathrm{Gal}(F/E)$ *is a solvable group, the extension* $E \subset F$ *is solvable.*

Let Ω be an algebraic closure of F and let K be the field generated in Ω by a primitive root of unity of order $[F : E]$. The extension $E \subset K$ is radical, Galois, and its Galois group is abelian. The extension $K \subset KF$ is Galois, and its Galois group is solvable, for it is a subgroup of $\mathrm{Gal}(F/E)$. Since $[KF : K]$ divides $[KF : E]$, K contains a primitive root of unity of order $[KF : K]$, it follows from c) that the extension $K \subset KF$ is radical. Therefore, the extension $E \subset KF$ is radical and the extension $E \subset F$ is solvable. □

Solving the "general equation of degree n" over some field K means being able to give formulae for solving any equation of degree n with arbitrary unspecified coefficients. In more precise terms, we want to solve the equation

$$X^n + a_1X^{n-1} + \cdots + a_n,$$

in which coefficients $a_1, \ldots, a_n$ are indeterminates. This is a polynomial equation over the field of rational functions $K(a_1, \ldots, a_n)$ in n indeterminates. By Exercise 3.11, its Galois group is equal to the full symmetric group $\mathfrak{S}_n$. Since this group is not solvable for $n \geqslant 5$ (Corollary 4.6.8), it follows from Theorem 5.6.3 that the general equation of degree n is not solvable by radicals, a result which had been first anticipated by the Italian mathematician P. Ruffini in 1799 and proved by N. Abel in 1826.

Corollary 5.6.5 (Abel). *Let K be a field. For $n \geqslant 5$, the general equation of degree n,*

$$X^n + a_1X^{n-1} + \cdots + a_n = 0,$$

viewed as a polynomial equation over the field $K(a_1, \ldots, a_n)$ of rational functions in n indeterminates and coefficients in K, is not solvable by radicals.

You will find below, and also in some exercises, explicit examples of polynomial equations (over the field of rational numbers) which are not solvable by radicals.

5.7 How (not) to compute Galois groups

In many actual applications, one considers a separable polynomial P, irreducible or not, with coefficients in a field K, and a splitting extension $K \subset L$ of the polynomial P, so that L is generated over K by the roots $x_1, \dots, x_n$ of P in an algebraic closure of K. As in Section 3.3, the Galois group $G = \mathrm{Gal}(L/K)$ is naturally a subgroup of the group of permutations of $\{x_1, \dots, x_n\}$, hence can be identified with a subgroup of the symmetric group $\mathfrak{S}_n$.

The first result of this section shows that, provided one knows how to factor polynomials with many indeterminates and coefficients in K, then one can explicitly determine the group G.

The group $G = \mathrm{Gal}(L/K)$ acts on the ring $L[Y_1, \dots, Y_n]$ coefficientwise, hence also on the the field of rational functions $L(Y_1, \dots, Y_n)$, which is its field of fractions. It also acts on the ring of polynomials $L[X, Y_1, \dots, Y_n]$. To simplify notation, we will write $\boldsymbol{Y}$ as an abbreviation for $Y_1, \dots, Y_n$. For example, we write $L[\boldsymbol{Y}]$ for $L[Y_1, \dots, Y_n]$ and $L(\boldsymbol{Y})$ for $L(Y_1, \dots, Y_n)$.

For any $\sigma \in \mathfrak{S}_n$, we let

$$\xi_\sigma = x_1 Y_{\sigma(1)} + \cdots + x_n Y_{\sigma(n)} \in L[\boldsymbol{Y}].$$

Lemma 5.7.1. a) *For any element τ in the Galois group* $\mathrm{Gal}(L/K)$, *one has*

$$\tau(\xi_\sigma) = \xi_{\sigma\tau^{-1}}.$$

b) *The extension $K(\boldsymbol{Y}) \subset L(\boldsymbol{Y})$ is Galois, with Galois group G.*

c) *Moreover, $\xi = x_1 Y_1 + \cdots + x_n Y_n$ is a primitive element.*

Proof. For any $\tau \in G$, one has

$$\tau(\xi_\sigma) = \sum_{i=1}^{n} \tau(x_i) Y_{\sigma(i)} = \sum_{i=1}^{n} x_{\tau(i)} Y_{\sigma(i)} = \sum_{j=1}^{n} x_j Y_{\sigma(\tau^{-1}(j))} = \xi_{\sigma\tau^{-1}},$$

which proves *a*).

b) If $R = P/Q \in L(Y)$, one can write

$$R = \frac{P}{Q} \prod_{\tau \in G\setminus\{1\}} \frac{\tau(Q)}{\tau(Q)} = \frac{P \prod_{\tau \neq 1} \tau(Q)}{\prod_{\tau} \tau(Q)},$$

a new fraction whose denominator D belongs to $K[\boldsymbol{Y}]$ since it is clearly invariant under any $\tau \in G$. Let $N = RD$ be its numerator, then R is invariant under G if and only if N is. It follows that $L(\boldsymbol{Y})^G = K(\boldsymbol{Y})$, and by Artin's lemma (Prop. 3.2.8), the extension $K(\boldsymbol{Y}) \subset L(\boldsymbol{Y})$ is Galois, with Galois group G.

c) Let $\xi = \xi_{\text{id}} = x_1Y_1 + \cdots + x_nY_n$. For any $\tau \in G$, $\tau(\xi) = \xi_{\tau^{-1}}$, so that $\tau = \text{id}$ is the only element of G such that $\tau(\xi) = \xi$. This shows that the extension $K(\boldsymbol{Y}) \subset L(\boldsymbol{Y})$ is generated by ξ. □

It follows from the Lemma that the minimal polynomial of ξ over $K(\boldsymbol{Y})$ is equal to

$$M_\xi(T) = \prod_{\tau \in G} (T - \tau(\xi)) = \prod_{\tau \in G} (T - \xi_\tau).$$

It belongs to $K[\boldsymbol{Y}, T]$ and is irreducible in $K(\boldsymbol{Y})[T]$, hence is irreducible in $K[\boldsymbol{Y}, T]$ for the ring $K[\boldsymbol{Y}]$ is a unique factorization domain.

Theorem 5.7.2. *Let us define a polynomial in variables $X, Y_1, \ldots, Y_n$ by the formula*

$$\mathscr{R}_P(T) = \prod_{\sigma \in \mathfrak{S}_n} (T - \xi_\sigma) = \prod_{\sigma \in \mathfrak{S}_n} (T - (x_1 Y_{\sigma(1)} + \cdots + x_n Y_{\sigma(n)})).$$

This is a separable polynomial with coefficients in K.

Let M be the unique irreducible factor of $\mathscr{R}_P$ in $K(\boldsymbol{Y})[T]$ which is monic in T and divisible by $T - \xi$ in $L(\boldsymbol{Y})[T]$.

Then $M = M_\xi$ and a permutation $\sigma \in \mathfrak{S}_n$ belongs to G if and only if

$$M(T, Y_1, \ldots, Y_n) = M(T, Y_{\sigma(1)}, \ldots, Y_{\sigma(n)}).$$

Proof. Any $\tau \in G$ induces a permutation of the roots of $\mathscr{R}_P$, since $\tau(\xi_\sigma) = \xi_{\sigma\tau^{-1}}$, hence $\tau(\mathscr{R}_P) = \mathscr{R}_P$ for any $\tau \in \text{Gal}(L/K)$ and the coefficients of $\mathscr{R}_P$ belong to K.

Since M and M_ξ have the common factor $X - \xi$ in $L(\boldsymbol{Y})[T]$, it follows from Corollary 2.4.3 that M and M_ξ have a common factor in $K(\boldsymbol{Y})[T]$. Being both monic and irreducible in $K(\boldsymbol{Y})[T]$, they are equal and

$$M(T, Y_1, \ldots, Y_n) = \prod_{\tau \in G} (T - (x_1 Y_{\tau(1)} + \cdots + x_n Y_{\tau(n)})).$$

Finally, for $\sigma \in \mathfrak{S}_n$, one has

$$\begin{aligned} M(T, Y_{\sigma(1)}, \ldots, Y_{\sigma(n)}) &= \prod_{\tau \in G} \big(X - x_1 Y_{\tau(\sigma(1))} - \cdots - x_n Y_{\tau(\sigma(n))}\big) \\ &= \prod_{\tau \in G\sigma} \big(X - x_1 Y_{\tau(1)} - \cdots - x_n Y_{\tau(n)}\big), \end{aligned}$$

so that

$$M(X, Y_{\sigma(1)}, \ldots, Y_{\sigma(n)}) = M(X, Y_1, \ldots, Y_n)$$

if and only if $G\sigma = G$, which means exactly that $\sigma \in G$. □

However nice it may look, this theorem is of almost no practical use. For example, if K is the field of rational numbers, the complexity of factoring multivariate polynomials of large degree ($\deg \mathscr{R}_P = n!$) is tremendous and this approach rapidly fails, even on the fastest available computing systems. We will still deduce from it a fundamental theoretical consequence concerning the behaviour of the Galois group of a polynomial by *specialization* of its coefficients, which is the subject of the next section.

Observe that the polynomial $\mathscr{R}_P$ defined in the theorem is symmetric in $Y_1, \dots, Y_n$, and is independent of the particular numbering of the roots. On the contrary, its irreducible factor M depends on the chosen numbering, as well as the Galois group, viewed as a subgroup of a symmetric group. Let us make this dependence explicit.

Let $P \in K[X]$ be a separable polynomial of degree n and let $K \to L$ be a splitting extension of P. Let R be the set of roots of P in L. A numbering of R is a bijection $\nu\colon \{1, \dots, n\} \xrightarrow{\sim} R$; it defines an embedding $\lambda_\nu\colon \operatorname{Gal}(L/K) \hookrightarrow \mathfrak{S}_n$ such that

$$\nu(\lambda_\nu(g)(i)) = g(\nu(i)), \quad g \in \operatorname{Gal}(L/K), \quad i \in \{1, \dots, n\}.$$

Denote its image by $G_\nu = \lambda_\nu(\operatorname{Gal}(L/K))$. Observe that the polynomial $\mathscr{R}_P$ satisfies

$$\begin{aligned}\mathscr{R}_P(T) &= \prod_{\sigma \in \mathfrak{S}_n} \bigl(T - (x_{\sigma^{-1}(1)} Y_1 + \dots + x_{\sigma^{-1}(n)} Y_n)\bigr) \\ &= \prod_{\text{numberings } \nu} \bigl(T - (\nu(1) Y_1 + \dots + \nu(n) Y_n)\bigr),\end{aligned}$$

the last product being over all numberings of the roots of P. Let $\mathscr{R}_{P,\nu}$ denote the minimal polynomial of $\xi = \nu(1)Y_1 + \dots + \nu(n)Y_n$ introduced above, so that

$$\begin{aligned}\mathscr{R}_{P,\nu}(T, Y_1, \dots, Y_n) &= \prod_{\tau \in G} \bigl(T - (\tau(\nu(1))Y_1 + \dots + \tau(\nu(n))Y_n)\bigr) \\ &= \prod_{\sigma \in G_\nu} \bigl(T - (\nu(\sigma(1))Y_1 + \dots + \nu(\sigma(n))Y_n)\bigr).\end{aligned}$$

If μ is another numbering, there is a permutation $\sigma \in \mathfrak{S}_n$ such that $\mu(i) = \nu(\sigma(i))$ for any $i \in \{1, \dots, n\}$. Then either $\mathscr{R}_{P,\mu}$ and $\mathscr{R}_{P,\nu}$ are coprime, or they have a factor in common. In this case, they are necessarily equal since they both are irreducible and monic; moreover, one has $\sigma \in G_\nu$. This implies that $\mathscr{R}_P$ is the product of $\mathscr{R}_{P,\nu\circ\sigma}$, the σ being some representatives in $\mathfrak{S}_n$ of all left cosets of G_ν. The embeddings of the Galois group into $\mathfrak{S}_n$ defined by μ and ν satisfy the relation

$$\lambda_\mu(g) = \sigma^{-1}\lambda_\nu(g)\sigma.$$

In particular, $G_\mu = \sigma^{-1}G_\nu\sigma$ is conjugate to G_ν in $\mathfrak{S}_n$.

5.8 Specializing Galois groups

Before I give a general definition, let me explain two important examples:

a) Let P be a monic polynomial with integer coefficients, and let G denote the Galois group of a splitting extension of P over $\mathbf{Q}$. For any prime number p, one can reduce the polynomial P modulo p; hence one obtains a new Galois group G_p corresponding to a finite extension of $\mathbf{Z}/p\mathbf{Z}$.

b) Let $P \in \mathbf{Q}(t)[X]$ be a polynomial with coefficients in the field $\mathbf{Q}(t)$ of rational functions, denote by G the Galois group of a splitting extension of P over $\mathbf{Q}(t)$. For any rational number α which is not a pole of the coefficients of P, one can evaluate the polynomial P at $t = \alpha$, and obtain a polynomial $P_\alpha \in \mathbf{Q}[X]$, hence a Galois group G_α.

We will see that the Galois groups of these specialized equations are, in a quite natural way, *subgroups* of the group G.

Definition 5.8.1. *A* place *of the field K is a map $\varphi\colon K \to k \cup \{\infty\}$, where k is a field, which satisfies the following properties:*

a) *if $\varphi(x)$ and $\varphi(y)$ are not both ∞, then $\varphi(x+y) = \varphi(x) + \varphi(y)$, with the convention $a + \infty = \infty$ for $a \in k$;*

b) *if $\{\varphi(x), \varphi(y)\} \neq \{0, \infty\}$, then $\varphi(xy) = \varphi(x)\varphi(y)$, with the convention $a\infty = \infty$ for $a \neq 0$.*

Example 5.8.2. Let us go back to the two previous examples.

a) Let p be a prime number. Let $x = a/b$ be a rational number, written in smallest terms. If p divides b, let us set $\varphi_p(x) = \infty$. If p does not divide b, let $\varphi_p(x)$ be the quotient in $\mathbf{Z}/p\mathbf{Z}$ of the classes of a and b modulo p. This map $\varphi_p\colon \mathbf{Q} \to (\mathbf{Z}/p\mathbf{Z}) \cup \{\infty\}$ is a place.

b) Let $\alpha \in \mathbf{Q}$. A rational function has a "value" at α, which is set to ∞ if α is a pole. This map $\varphi_\alpha\colon \mathbf{Q}(t) \to \mathbf{Q} \cup \{\infty\}$ is a place.

If $\varphi\colon K \to k \cup \{\infty\}$ is a place, let $A = \varphi^{-1}(K)$ be the set of $x \in K$ such that $\varphi(x) \neq \infty$. The definition of a place implies at once that A is a subring of K, which we will call the *valuation ring of* φ. (*Exercise:* check it! See also Exercise 5.15.) In the two examples above, any ideal in A is generated by a power of p, or of $X - \alpha$, accordingly. In particular, *in these two cases, the ring A is a principal ideal ring.*

Let us fix a place $\varphi\colon K \to k \cup \{\infty\}$ of the field K. Let A denote the valuation ring of φ. Let $P \in K[X]$ be a monic polynomial of degree n. Assume that $P \in A[X]$, and that its discriminant $\Delta \in A$ satisfies $\varphi(\Delta) \neq 0$, so that the polynomial $\varphi(P) \in k[X]$ is separable. Let G be the Galois group of a splitting extension L of P over K, and let H be the Galois group of a splitting extension ℓ of the polynomial $\varphi(P)$ over k.

Lemma 5.8.3. *The polynomial $\mathscr{R}_P$ belongs to $A[T, \mathbf{Y}]$, and $\mathscr{R}_{\varphi(P)} = \varphi(\mathscr{R}_P)$.*

Proof. Let us first consider the polynomial

$$R = \prod_{\sigma \in \mathfrak{S}_n} (T - (\sum_{i=1}^{n} X_{\sigma(i)} Y_i)).$$

We view it as a polynomial in $T, Y_1, \dots, Y_n$ with coefficients in $\mathbf{Z}[X_1, \dots, X_n]$, writing

$$R = \sum_{I=(i_0,\dots,i_n)} R_I(X_1, \dots, X_n) Y^{i_0} Y_1^{i_1} \dots Y_n^{i_n}.$$

The polynomial R is symmetric in $X_1, \dots, X_n$, hence each of its coefficients R_I is symmetric too. Therefore, there is for each I a polynomial $\tilde{R}_I \in \mathbf{Z}[S_1, \dots, S_n]$ such that

$$R_I(X_1, \dots, X_n) = \tilde{R}_I(S_1(X), \dots, S_n(X)).$$

Let us write $P = X^n + a_1 X^{n-1} + \dots + a_n$ and let $x_1, \dots, x_n$ denote the roots of P in L, so that $a_j = (-1)^j S_j(x_1, \dots, x_n)$. It follows that

$$\mathscr{R}_P = \sum_I \tilde{R}_I(-a_1, \dots, (-1)^n a_n) T^{i_0} Y_1^{i_1} \dots Y_n^{i_n}.$$

Since the coefficients a_j belong to the subring A, $\mathscr{R}_P \in A[T, \mathbf{Y}]$.

Moreover, one has $\varphi(P) = X^n + \varphi(a_1) X^{n-1} + \dots + \varphi(a_n)$, and the same argument shows that

$$\tilde{R}_{\varphi(P)} = \sum_I \tilde{R}_I(-\varphi(a_1), \dots, (-1)^n \varphi(a_n)) T^{i_0} Y_1^{i_1} \dots Y_n^{i_n}.$$

Consequently,

$$\varphi(R_P) = \sum_I \varphi(\tilde{R}_I(-a_1, \dots, (-1)^n a_n)) T^{i_0} Y_1^{i_1} \dots Y_n^{i_n} = \tilde{R}_{\varphi(P)}$$

is the polynomial attached to $\varphi(P)$, which proves the lemma. □

Lemma 5.8.4. *For any numbering ν of the roots of P in L, the polynomial $\mathscr{R}_{P,\nu}$ belongs to $A[T, \mathbf{Y}]$.*

Proof. If the ring A is a unique factorization domain, e.g. , in the two examples above, it follows from Theorem 2.4.7 that the polynomial $\mathscr{R}_{P,\nu}$ belongs to $A[T, \boldsymbol{Y}]$. That remains true in the general case, for "a valuation ring is integrally closed," but we shall not prove it here; see Exercise 5.16. □

We saw that the irreducible factors in $k[T, \boldsymbol{Y}]$ of the polynomial $\mathscr{R}_{\varphi(P)}$ were of the form $\mathscr{R}_{\varphi(P),\mu}$ for μ a numbering of the roots of $\varphi(P)$ in ℓ. Now, since $\mathscr{R}_{P,\nu}$ divides $\mathscr{R}_P$, the preceding lemmas show that $\varphi(\mathscr{R}_{P,\nu})$ is a divisor of $\mathscr{R}_{\varphi(P)}$ in $k[T, \boldsymbol{Y}]$. We shall say that a numbering ν of the roots of P and a numbering μ of the roots of $\varphi(P)$ are *compatible* if $\mathscr{R}_{\varphi(P),\mu}$ divides $\varphi(\mathscr{R}_{P,\nu})$.

Theorem 5.8.5. *Fix a numbering ν of the roots of P, hence an embedding $\lambda_\nu\colon \mathrm{Gal}(L/K) \to \mathfrak{S}_n$ of image G_ν.*

a) *There exists a numbering μ of the roots of $\varphi(P)$ which is compatible with ν. It defines an embedding of the Galois group H into $\mathfrak{S}_n$; its image H_μ is a subgroup of G_ν.*

b) *Let μ' be any numbering of the roots of $\varphi(P)$, and let σ be the unique permutation $\in \mathfrak{S}_n$ such that $\mu'(i) = \mu(\sigma(i))$ for any $i \in \{1, \dots, n\}$. Then μ' is compatible with ν if and only if $\sigma \in G_\nu$. In that case, $H_{\mu'} = \sigma^{-1} H_\mu \sigma$ is conjugate to H_μ in G_ν.*

This shows that "the" Galois group H of the specialized equation $\varphi(P)$ is in an almost natural way a *subgroup* of the Galois group G of the equation P. Moreover, if the group G is abelian, or if the group H appears to be normal in G, then the Galois group of the specialized equation is a *canonical* subgroup of the Galois group.

Proof. The irreducible factors of the polynomial $\varphi(\mathscr{R}_{P,\nu}) \in k[T, \boldsymbol{Y}]$ divide $\mathscr{R}_{\varphi(P)}$, hence are of the form $\mathscr{R}_{\varphi(P),\mu}$ for some numberings μ of the roots of $\varphi(P)$ in ℓ. These numberings are precisely those which are compatible with ν.

More precisely, with N denoting the set of numberings of the roots of $\varphi(P)$ which are compatible with ν, one has the formula

$$\varphi(\mathscr{R}_{P,\nu}) = \prod_{\mu \in N} (T - (\mu(1)Y_1 + \cdots + \mu(n)Y_n))$$

in $\ell[T, \boldsymbol{Y}]$. Let $\sigma \in G_\nu$; then

$$\mathscr{R}_{P,\nu}(T, Y_{\sigma(1)}, \dots, Y_{\sigma(n)}) = \mathscr{R}_{P,\nu}(T, Y_1, \dots, Y_n),$$

hence, taking the images of both sides by φ,

$$\prod_{\mu \in N} (T - (\mu(1)Y_{\sigma(1)} + \cdots + \mu(n)Y_{\sigma(n)})) = \prod_{\mu \in N} (T - (\mu(1)Y_1 + \cdots + \mu(n)Y_n)).$$

Writing $\mu(i) = \mu \circ \sigma^{-1}(\sigma(i))$, we find that $N\sigma^{-1} = N$; in other words, $N = NG_\nu$ is a right coset modulo G_ν. Since the cardinality of N is that of G_ν, one has $N = \mu G_\nu$ for any $\mu \in N$.

Fix such a μ. The polynomial $\mathscr{R}_{\varphi(P),\mu}$ divides $\varphi(R_{P,\nu})$. Looking in $\ell[T, \mathbf{Y}]$, one sees that $\mu H_\mu \subset N = \mu G_\nu$. Consequently, $H_\mu \subset G_\nu$.

If μ' is another numbering, one has $\mu' = \mu \circ \sigma$ for some $\sigma \in \mathfrak{S}_n$. Moreover, μ' is compatible with ν if and only if $\mu' \in N$, hence if and only if $\sigma \in G_\nu$. For such a numbering μ', we saw that $H_{\mu'} = \sigma^{-1} H_\mu \sigma$. The subgroups $H_{\mu'}$ and H_μ are therefore conjugate in G_ν. □

Let me now show some examples of how this theorem can be used to specify the shape of the Galois group of a polynomial with rational coefficients. Recall a remark from the end of Section 3.5 on finite fields. We define the *shape* of a permutation of $\{1, \dots, n\}$ as the partition of n that it defines (see p. 93).

Lemma 5.8.6. *Let P be a monic separable polynomial with coefficients in a finite field k. Let us denote by $n_1, \dots, n_r$ the degrees of the irreducible factors of P in $k[X]$. Let $k \to \ell$ be a splitting extension of P; the Frobenius automorphism $F \in \operatorname{Gal}(\ell/k)$ induces a permutation of the roots of P in ℓ. This permutation has shape $(n_1, \dots, n_r)$.*

Recall also from Prop. 4.6.1 that the conjugacy class of this permutation is characterized by these integers $(n_1, \dots, n_r)$. Consequently, this lemma and Theorem 5.8.5 allow one to exhibit *conjugacy classes* of elements in the Galois group. In some cases, this is even enough to compute the Galois group!

Example 5.8.7. Let us begin with the polynomial $P = X^5 - X - 1$. Denote by G its Galois group over $\mathbf{Q}$, considered as a subgroup of the group of permutations of the 5 roots, identified with $\mathfrak{S}_5$.

Reducing the polynomial modulo 2, we see that it has no root in $\mathbf{F}_2$, but it has two in $\mathbf{F}_4$. Indeed, the g.c.d. of $X^5 - X - 1$ and $X^4 - X$ is equal to $X^2 - X - 1$ over $\mathbf{F}_2$, so that $P \pmod 2$ has a factor of degree 2, the other being necessarily of degree 3. In particular, $P \pmod 2$ is separable over $\mathbf{F}_2$ and its Galois group over $\mathbf{F}_2$ is generated by an element of $\mathfrak{S}_5$ of shape $(2, 3)$. By Theorem 5.8.5, G contains a permutation of this shape, hence its cube, which is a transposition.

Let us now reduce modulo 3. By computing the g.c.d. of $P \pmod 3$ and $X^3 - X$, *resp.* $X^9 - X$ (computer algebra systems can be of great use in such calculations...), we check that $P \pmod 3$ has no root in $\mathbf{F}_3$, nor in $\mathbf{F}_9$. (*Exercise:* do it also by hand, using, for example, the fact that for any element $x \in \mathbf{F}_9$, one has $x^4 \in \{0, \pm 1\}$.) It follows that $P \pmod 3$ is irreducible over $\mathbf{F}_3$. By Theorem 5.8.5, G contains a 5-cycle. Incidentally, this shows that the polynomial P is irreducible.

It now follows from Proposition 4.6.2 that G is equal to the full symmetric group $\mathfrak{S}_5$. By the way, this gives an explicit example of a polynomial with rational coefficients which cannot be solved by radicals, for its Galois group, being $\mathfrak{S}_5$, is not solvable.

Example 5.8.8. Let us show in a similar way that the Galois group G of the polynomial $P = X^5+20X-16$ over $\mathbf{Q}$, viewed as a subgroup of $\mathfrak{S}_5$, is equal to the alternating group $\mathfrak{A}_5$. Modulo 2, one has $P \equiv X^5$, which is not separable. Let us thus look modulo 3. One has $P \equiv X^5 - X - 1 \pmod 3$; as we saw in the previous example, $P \pmod 3$ is irreducible. As above, the group G contains a 5-cycle.

Modulo 7, one has $P \equiv X^5 - X - 2$ and its roots in $\mathbf{F}_7$ are 2 and 3; moreover, one has

$$P \equiv (X-2)(X-3)(X^3-2X^2-2X+2) \pmod 7.$$

The polynomial $X^3 - 2X^2 - 2X + 2$ has no root in $\mathbf{F}_7$ (check it!), hence is irreducible since its degree is 3. It follows that G contains a 3-cycle.

Modulo 23, one gets a factorization of P as the product of a linear factor and two polynomials of degree 2, hence there is a permutation of the form $(1)(2,3)(4,5)$ — a double transposition — in G.

Considering other prime numbers does not seem to give new information on G. We already know that the order of G is a multiple of 2, 3 and 5, hence of their l.c.m. 60, and since it is a subgroup of $\mathfrak{S}_5$, its order divides by $5! = 120$.

We now have to use another piece of information. Observe that the discriminant of P is equal to

$$5^5 \times (-16)^4 + 4^4 \times 20^5 = 1024000000 = 2^{16}\, 5^6 = (2^8\, 5^3)^2$$

(see Exercise 3.22), so is a square in $\mathbf{Q}$. By Proposition 3.4.2, this implies that G is a subgroup of $\mathfrak{A}_5$. Since $\operatorname{card} \mathfrak{A}_5 = 60$, one necessarily has $G = \mathfrak{A}_5$.

In more complicated examples, these two ingredients, reduction modulo prime numbers and the consideration of the discriminant, are not enough and one is forced to use more general resolvent polynomials (see Section 3.4).

Example 5.8.9. Computer algebra systems like MAGMA, PARI/GP or MAPLE can compute Galois groups for you, at least if the degree is not too big. For instance, here is the output of a (verbose) MAPLE session when asked to compute the Galois group of the polynomial $t^5 - 5t + 12$ over the rationals.

```
> infolevel[galois]:=2;
> galois(t^5-5*t+12);
galois:   Computing the Galois group of    t^5-5*t+12
```

```
galois/absres:   64000000 = ``(8000)^2
galois/absres:   Possible groups:   {"5T2", "5T1", "5T4"}
galois/absres:   p = 3    gives shape   2, 2, 1
galois/absres:   Removing   {"5T1"}
galois/absres:   Possible groups left:   {"5T2", "5T4"}
galois/absres:   p = 7    gives shape   5
galois/absres:   p = 11   gives shape   5
galois/absres:   p = 13   gives shape   5
galois/absres:   p = 17   gives shape   2, 2, 1
galois/absres:   p = 19   gives shape   5
galois/absres:   p = 23   gives shape   5
galois/absres:   p = 29   gives shape   2, 2, 1
galois/absres:   p = 31   gives shape   2, 2, 1
galois/absres:   p = 37   gives shape   5
galois/absres:   p = 41   gives shape   5
galois/absres:   The Galois group is probably one of   {"5T2"}
galois/respol:   Using the orbit-length partition of 2-sets.
galois/respol:   Calculating a resolvent polynomial...
galois/respol:   Factoring the resolvent polynomial...
galois/respol:   Orbit-length partition is   5, 5
galois/respol:   Removing   {"5T4"}
galois/respol:   Possible groups left:   {"5T2"}
        "5T2", {"5:2", "D(5)"}, "+", 10, {"(1 4)(2 3)", "(1 2 3 4 5)"}
```

To understand these lines, one needs to know that, up to conjugacy, there are only 5 transitive subgroups of $\mathfrak{S}_5$. These are

a) the cyclic group C_5, generated by the 5-cycle $(1,2,3,4,5)$, isomorphic to $\mathbf{Z}/5\mathbf{Z}$ and denoted in this context as 5T1;

b) the dihedral group D_5, generated by $(1,2,3,4,5)$ and $(2,5)(3,4)$, denoted as 5T2;

c) the metacyclic group M_{20}, defined as the normalizer 5T3 of C_5 in $\mathfrak{S}_5$, of cardinality 20, also isomorphic to the group of all maps $\mathbf{F}_5 \to \mathbf{F}_5$ of the form $x \mapsto ax+b$ with $a \in \mathbf{F}_5^*$ and $b \in \mathbf{F}_5$;

d) the alternating group $\mathfrak{A}_5$, of cardinality 60 and denoted 5T4;

e) the full symmetric group $\mathfrak{S}_5$, denoted 5T5.

(In fact, all practical algorithms for computing Galois groups require the list of all transitive subgroups of $\mathfrak{S}_n$, which is known up to $n = 31$. The notations 5T1, etc. come from this classification.)

First, the discriminant is computed. It is a square, $(64,000,000 = (8000)^2)$, hence the group must be a subgroup of the alternating group, which excludes M_{20} and $\mathfrak{S}_5$ (respectively 5T3 and 5T5). Then, the program reduces our polynomial modulo small prime numbers and computes its factorization over the corresponding finite field, hence the shape of some permutation belonging to the Galois group; then, for any group which has not yet been excluded, the program simply checks whether it contains such a permutation. In fact, all nontrivial elements of the group generated by a 5-cycle are 5-cycles themselves, so that the group C_5 (5T1) is eliminated at once by reducing modulo $p = 3$. However, no new information is obtained in this way by reducing modulo prime numbers $\leqslant 41$.

Then, MAPLE indicates that the group would probably be equal to D_5 (5T2). Indeed, by Chebotarëv's density theorem, a profound and difficult theorem from algebraic number theory, all possible conjugacy classes of elements in the Galois group will appear by reducing modulo larger and larger prime numbers, and they will appear "in proportion" to their cardinalities. In fact, the shape of a permutation detects only its conjugacy class in the symmetric group, so that an easier result, due to Frobenius, is sufficient for our purposes. The number of permutations of a given shape in each group is given in Table 5.1. In our example, the shapes that appear are $(2, 2)$, 4 times, and (5), 7 times. If the group had been $\mathfrak{A}_5$ (5T4), the shape (3) would probably have already appeared, therefore MAPLE suggests that the group is D_5.

	C_5 (5T1)	D_5 (5T2)	M_{20} (5T3)	$\mathfrak{A}_5$ (5T4)	$\mathfrak{S}_5$ (5T5)
1,1,1,1,1	1	1	1	1	1
2,1,1,1					10
3,1,1				20	20
2,2,1		5	5	15	15
4,1			10		30
3,2					20
5	4	4	4	24	24
total	5	10	20	60	120

Table 5.1. Number of permutations inducing a given partition in subgroups of $\mathfrak{S}_5$

Since D_5 is a subgroup of $\mathfrak{A}_5$, it remains to check whether G is, up to conjugacy, a subgroup of D_5. This requires a resolvent polynomial like

$$X_1X_2 + X_2X_3 + X_3X_4 + X_4X_5 + X_5X_1,$$

whose stabilizer is exactly D_5. (Can you see why? Remember that D_5 is the symmetry group of the regular pentagon.) Computing the complex roots of the polynomial $t^5 - 5t + 12$ with large accuracy, one can evaluate the above resolvent polynomial at all permutations of the roots. Some of these evaluations are integers and Prop. 3.4.5 implies that the Galois group is equal to D_5. In fact, a floating point calculation does not really *prove* that the numbers obtained are integers, only that they are up to the given precision. However, using results such as Liouville's theorem (Exercise 1.2), one can prove that the numbers obtained are actually integers.

5.9 Hilbert's irreducibility theorem

This section explains some facts concerning the variation of the Galois group of a polynomial equation depending on a parameter. Any of the three theorems below constitute what is generally known as Hilbert's irreducibility theorem.

Let us consider a monic polynomial P with coefficients in the field $\mathbf{Q}(T)$ of rational functions. Let us assume that P is irreducible as a polynomial in $\mathbf{Q}(T)[X]$. We will first show that for "many" values $t \in \mathbf{Z}$, the polynomial $P(t, X) \in \mathbf{Q}[X]$ has no root in $\mathbf{Q}$. We will then show that in fact, for "many" integers t, the polynomial $P(t, X)$ is even irreducible. Recall from Theorem 5.8.5 that, essentially, the Galois group over $\mathbf{Q}$ of the polynomial $P(t, X)$ is a subgroup of the Galois group over $\mathbf{Q}(T)$ of the polynomial $P(T, X)$. The last result, Theorem 5.9.7, claims that for "many" integers t, these two groups are in fact *equal!*

This is a theorem in *arithmetic*, as opposed to algebra, and it relies on properties of the field $\mathbf{Q}$ of rational numbers. It is obviously false if one replaces $\mathbf{Q}(T)$ by $\mathbf{C}(T)$ in its statement: there are irreducible polynomials $P \in \mathbf{C}(T)[X]$ of any degree but for any t, the polynomial $P(t, X)$ is split in $\mathbf{C}$, for the field of complex numbers is algebraically closed. The Galois group of the specialized equation is therefore trivial.

The heart of the arithmetic arguments will be in the proof of Prop. 5.9.1, at the point when we bound from below by 1 the absolute value of a nonzero integer. Remark that such a lower bound was also the crucial point in the proof that e and π are transcendental numbers (Theorems 1.6.3 and 1.6.6). However, the arguments we will use to prove Theorems 5.9.4, 5.9.6 and 5.9.7 from that proposition are essentially of algebraic nature.

Proposition 5.9.1. *Let e be any positive integer and let $\varphi = \sum_{n \geqslant -n_0} a_n u^{-n/e}$ be a Laurent series in the variable $u^{-1/e}$ which is not a polynomial in u. (In other words, there is a nonzero coefficient a_n such that either $n > 0$ or e does not divide n.) Assume that $\varphi(u)$ converges for $|u| \geqslant B_0$. Denote by $N(B)$ the number of integers $u \in [B_0, B]$ such that $\varphi(u) \in \mathbf{Z}$. Then, there exists a real number $\alpha < 1$ such that $N(B)/B^\alpha$ remains bounded when $B \to \infty$.*

From now on, we shall use the big-O notation and write $N(B) = \mathrm{O}(B^\alpha)$ to mean that $N(B)/B^\alpha$ remains bounded when $B \to \infty$.

Proof. It suffices to separately consider the real and imaginary parts of φ, for at least one of them is not a polynomial. We will therefore assume that φ has real coefficients. Observe that φ defines a $\mathscr{C}^\infty$ function from the interval $(B_0, +\infty)$ to $\mathbf{R}$, its derivatives of any order being obtained by deriving the series term by term. Hence, for $m > n_0/e$, $\varphi^{(m)}(u)$ decreases to 0 when $u \to +\infty$ Since φ is not a polynomial, $\varphi^{(m)}$ is not the zero-function and, when $u \to \infty$, $\varphi^{(m)}(u)$ is then equivalent to its first term, which is of the form $cu^{-\mu}$ for some real number $c \neq 0$ and some positive real number μ. In particular, for u large enough, say $u \geqslant B_1$, one has an inequality $c_1 u^{-\mu} \leqslant \left|\varphi^{(m)}(u)\right| \leqslant c_2 u^{-\mu}$.

Let S denote the set of integers $\geqslant B_0$ such that $\varphi(u) \in \mathbf{Z}$. Consider $m+1$ elements in S, $u_0 < \cdots < u_m$, with $u_0 > B_1$ and let us introduce the determinant

$$D = \begin{vmatrix} 1 & \dots & 1 \\ u_0 & \dots & u_m \\ \vdots & & \vdots \\ u_0^{m-1} & \dots & u_m^m \\ \varphi(u_0) & \dots & \varphi(u_m) \end{vmatrix}.$$

This determinant is an integer, for it is the determinant of a matrix with integer coefficients. By Lemma 5.9.3 below, there exists a real number $\xi \in (u_0, u_m)$ such that

$$D = \frac{1}{m!}\varphi^{(m)}(\xi) \prod_{i>j} (u_i - u_j).$$

Since $u_0 \geqslant B_1$, $\varphi^{(m)}(\xi) \neq 0$; in particular $D \neq 0$. Since D is an integer, one has $|D| \geqslant 1$, hence a lower bound

$$\prod_{i>j} (u_i - u_j) \geqslant \frac{m!}{\left|\varphi^{(m)}(\xi)\right|} \geqslant \frac{m!}{c_2} \xi^\mu,$$

and, *a fortiori*,

$$(u_m - u_0)^{m(m+1)/2} \geqslant \frac{m!}{c_2} u_0^\mu.$$

We thus have shown the existence of positive real numbers b and β such that, for any $m+1$ elements $u_0 < \cdots < u_m$ in S with $u_0 > B_1$, one has

$$u_m \geqslant u_0 + b u_0^\beta. \tag{5.9.2}$$

Now we set $\alpha = 1/(1+\beta)$ and we split the interval $[B_0, B]$ as $[B_0, B^\alpha] \cup [B^\alpha, B]$. The interval $[B_0, B^\alpha]$ contains at most B^α elements of S. For B large enough, $B^\alpha \geqslant B_1$ and the lower bound (5.9.2) implies that the interval $[B^\alpha, B]$ contains at most $(m/b)B^{1-\alpha\beta} = (m/b)B^\alpha$ elements of S. Finally, for $B \geqslant B_1^{1/\alpha}$, $N(B) \leqslant (1 + m/b)B^\alpha$, as we had to prove. □

Lemma 5.9.3. *Let I be an interval in $\mathbf{R}$, and $f\colon I \to \mathbf{R}$ a function with $\mathscr{C}^n$-regularity. Let $x_0, \dots, x_n$ be elements in I. Then, there is $\xi \in I$ such that*

$$\begin{vmatrix} 1 & \dots & 1 \\ x_0 & \dots & x_n \\ \vdots & & \vdots \\ x_0^{n-1} & \dots & x_n^{n-1} \\ f(x_0) & \dots & f(x_n) \end{vmatrix} = \frac{f^{(n)}(\xi)}{n!} \prod_{i>j} (x_i - x_j).$$

Proof. It suffices to consider the case where all x_i are distinct. Let us consider x_0 as a parameter and denote by $D(x_0)$ the determinant above. For $A \in \mathbf{R}$, let $F_A\colon I \to \mathbf{R}$ be the function defined by $F_A(x) = D(x) - A \prod_{i=1}^{n} (x - x_i)$. This function F_A vanishes at $x_1, \dots, x_n$; let us choose A so that it vanishes at $x = x_0$ too.

By Rolle's Lemma, the derivative of F_A vanishes at n distinct points on I, then its second derivative $(n-1)$ times, and so on. Finally, there is at least one $\xi \in I$ such that $F_A^{(n)}(\xi) = 0$. Moreover,

$$\begin{aligned} F_A^{(n)}(\xi) = D^{(n)}(\xi) - An! &= \begin{vmatrix} 0 & 1 & \dots & 1 \\ x_0 & \dots & x_n & \\ \vdots & \vdots & & \vdots \\ 0 & x_1^{n-1} & \dots & x_n^{n-1} \\ f^{(n)}(\xi) & f(x_1) & \dots & f(x_n) \end{vmatrix} - An! \\ &= (-1)^n f^{(n)}(\xi) \begin{vmatrix} 1 & \dots & 1 \\ x_0 & \dots & x_n \\ \vdots & & \vdots \\ x_1^{n-1} & \dots & x_n^{n-1} \end{vmatrix} - An!, \end{aligned}$$

hence $A = (-1)^n \dfrac{f^{(n)}(\xi)}{n!} \prod_{i>j\geqslant 1} (x_i - x_j)$ and

$$D(x_0) = A \prod_{i=1}^{n} (x_0 - x_i) = \frac{f^{(n)}(\xi)}{n!} \prod_{i>j} (x_i - x_j).$$

This proves the lemma. □

Theorem 5.9.4. *Let P be a monic polynomial in $\mathbf{Q}(T)[X]$. Let $N(B)$ denote the number of integers $t \in [0, B]$ such that $P(t, X)$ has a root in $\mathbf{Q}$. If P has no root in $\mathbf{Q}(T)$, then there is a real number $\alpha < 1$ such that, when $B \to \infty$, $N(B) = \mathrm{O}(B^\alpha)$.*

Lemma 5.9.5. *Let n denote the degree of P. There exist an integer $e \geqslant 1$, Laurent series $x_1, \dots, x_n$ with complex coefficients, and a nonzero radius of convergence, such that for any complex number t of large enough modulus, the n complex roots of $P(t^e, X)$ are the $x_j(1/t)$, for $1 \leqslant j \leqslant n$.*

Proof. Since we look at the roots of $P(t, X)$ for t large, let us make a change of variables $t = 1/u$. Let R denote a common denominator to the coefficients of the polynomial $P(1/U, X) \in \mathbf{Q}(U)[X]$, so that $R(U)P(1/U, X) \in \mathbf{Q}[U, X]$. Multiplying by $R(U)^{n-1}$, we can then find a polynomial $Q \in \mathbf{Q}[U, Y]$, monic and of degree n with respect to Y, such that $P(1/U, X)R(U)^n = Q(U, R(U)X)$. By Puiseux's theorem (Theorem 2.6.1), there are power series $y_1, \dots, y_n$ with positive radius of convergence, and an integer $e \geqslant 1$ such that, for $|u|$ small enough, the roots of the polynomial $Q(u^e, Y)$ are the $y_j(u)$, for $1 \leqslant j \leqslant n$. Let us set $x_j(u) = R(u)^{-e} y_j(u)$. Expanding $R(u)^{-e}$ as a Laurent series around $u = 0$, one sees that the x_j are Laurent series, converging for $|u|$ small enough, but $u \neq 0$. Making the change of variables $t = 1/u$ again, the $x_j(1/t)$ are the roots of $P(t^e, X)$ provided $|t|$ is large enough. □

Proof of Theorem 5.9.4. Let $D \in \mathbf{Z}[T]$ be a common denominator of the coefficients of P, so that $P(T, X)D(T) \in \mathbf{Z}[T, X]$. There is a polynomial $Q \in \mathbf{Z}[T, X]$, monic as a polynomial in X, such that $P(T, X)D(T)^n = Q(T, D(T)X)$. The polynomial Q has no root in $\mathbf{Q}(T)$ (if $R(T)$ were a root of Q in $\mathbf{Q}(T)$, then $R(T)/D(T)$ would be a root of P in $\mathbf{Q}(T)$). Similarly, if $D(t) \neq 0$, then the polynomial $P(t, X) \in \mathbf{Z}[X]$ has a root in $\mathbf{Q}$ if and only if $Q(t, X)$ has a root in $\mathbf{Q}$. Therefore, it suffices to prove the theorem for the polynomial Q, which allows us to assume that $P \in \mathbf{Z}[T, X]$. Then, for any $t \in \mathbf{Z}$, the polynomial $P(t, X)$ is monic with integer coefficients. By Exercise 1.5, its roots in $\mathbf{Q}$ are necessarily integers.

Let $x_1, \dots, x_n$ be the series given by Lemma 5.9.5. Since P has no root in $\mathbf{Q}(T)$, none of these series is a polynomial. It is now enough to apply Proposition 5.9.1 to each of them and to add up the upper bounds obtained, so that we get the desired upper bound for $N(B)$. □

Theorem 5.9.6. *Let $P \in \mathbf{Q}(T)[X]$ be any monic irreducible polynomial with coefficients in $\mathbf{Q}(T)$. Let $N(B)$ denote the cardinality of the set of integers $t \in [0, B]$ such that t is not a pole of any coefficient of P and such that $P(t, X)$ is reducible in $\mathbf{Q}[X]$. Then there exists $\alpha < 1$ such that $N(B) = \mathrm{O}(B^\alpha)$.*

Proof. As in the proof of the preceding theorem, we assume that P belongs to $\mathbf{Z}[T, X]$. Let $x_1, \dots, x_n$ be the Laurent series given by Lemma 5.9.5. If t is large enough, say $t \geqslant B_0$, any monic factor of $P(t, X) \in \mathbf{Z}[X]$ has the form

$$P_I(t) = \prod_{i \in I} (X - x_i(t^{-1/e})),$$

where I is a subset of $\{1, \dots, n\}$. If $I \neq \emptyset$ and $I \neq \{1, \dots, n\}$, it is thus enough to show that the set of all integers $t \in [B_0, B]$ such that $P_I(t)$ belongs to $\mathbf{Z}[X]$ has cardinality $\mathrm{O}(B^\alpha)$.

But we may view the polynomial P_I as a polynomial with coefficients in the field K of converging Laurent series in a variable $T^{-1/e}$, and P_I is a factor of P in $K[X]$. Since P is irreducible in $\mathbf{Q}[T, X]$, the polynomial P_I does not belong to $\mathbf{Q}(T)[X]$ and at least one of its coefficients, say φ_I, is not a polynomial in T. Proposition 5.9.1 then implies that the set of all integers $t \in [B_0, B]$ such that $\varphi_I(t)$ is an integer has cardinality $\mathrm{O}(B^\alpha)$ for some $\alpha < 1$. The theorem is then proved. □

More generally, the following theorem says that the Galois group over $\mathbf{Q}$ of the polynomial $P(t, X)$, with $t \in \mathbf{Z}$, quite often coincides with the Galois group over $\mathbf{Q}(T)$ of the polynomial $P(T, X)$.

Theorem 5.9.7. *Let $P \in \mathbf{Q}(T)[X]$ be a monic polynomial with coefficients in $\mathbf{Q}(T)$. Let G denote its Galois group over $\mathbf{Q}(T)$. Let $N(B)$ be the cardinality of the set of all integers $t \in [0, B]$ such that either t is a pole of $P(T, X)$ or the Galois group of the polynomial $P(t, X)$ over $\mathbf{Q}$ is not isomorphic to G. Then, there exists $\alpha < 1$ such that $N(B) = \mathrm{O}(B^\alpha)$.*

Proof. As in the proof of Theorem 5.9.7, we assume that the coefficients of P are polynomials in T. Let us denote by n the degree of P in the variable X. Let $\mathbf{Q}(T) \to K$ be a splitting extension of the polynomial P and let $\kappa \in K$ be any primitive element. If $N = \operatorname{card} G$, then $N = [K : \mathbf{Q}(T)]$, and N is the degree of the minimal polynomial Q of κ over $\mathbf{Q}(T)$. The coefficients of Q are a priori rational functions in T. However, denoting by $D \in \mathbf{Q}[T]$ a common denominator of its coefficients, the minimal polynomial of $D(T)\kappa$ is equal to the polynomial $D(T)^N Q(T, D(T)^{-1}X)$ and therefore belongs to $\mathbf{Q}[T, X]$. This allows us to assume that $Q \in \mathbf{Q}[T, X]$.

Over $\mathbf{Q}(T)$, the polynomials P and Q have a common splitting extension, hence have the same Galois group, even if, as permutation groups, they look distinct (they do not act on the same set).

By the following lemma, there is a finite subset $S \subset \mathbf{Q}$ such that for any $t \notin S$, the polynomials $Q(t, X)$ and $P(t, X)$ are separable and have a common splitting extension $\mathbf{Q} \subset K_t$. By Theorem 5.8.5, the Galois group $\mathrm{Gal}(K_t/\mathbf{Q})$ can be considered as a subgroup of the Galois group $\mathrm{Gal}(K/\mathbf{Q}(T))$, so that $[K_t : \mathbf{Q}] \leqslant [K : \mathbf{Q}(T)] = N$. By Theorem 5.9.6 applied to the polynomial Q, there exists $\alpha < 1$ such that the number $N(B)$ of all integers $t \in [0, B]$ such that $t \notin S$ and such that $Q(t, X)$ is irreducible in $\mathbf{Q}[X]$, satisfies $N(B) = \mathrm{O}(B^\alpha)$. For such t, $[K_t : \mathbf{Q}] \geqslant N$, so that one has $[K_t : \mathbf{Q}] = N$ and $\mathrm{Gal}(K_t/\mathbf{Q})$ is isomorphic to $\mathrm{Gal}(K/\mathbf{Q}(T))$. □

Lemma 5.9.8. *Let $P \in \mathbf{Q}(T)[X]$ be a monic polynomial, let $\mathbf{Q}(T) \subset K$ be a splitting extension of P. Let $y \in K$ be a primitive element and denote by $Q \in \mathbf{Q}(T)[X]$ its minimal polynomial. There exists a finite subset $\Sigma \subset \mathbf{Q}$ such that for any $t \notin \Sigma$, the polynomials $Q(t, X)$ and $P(t, X)$ are separable and have a common splitting extension.*

Proof. Let us denote by $x_1, \dots, x_n$ the roots of P in K. One can find polynomials $A_i \in \mathbf{Q}(T)[Y]$ such that for any i, $x_i = A_i(y)$, in other words,

$$P(T, X) = \prod_{i=1}^{n} (X - A_i(T, y)).$$

Replacing y by a formal variable Y, this implies that $Q(T, Y)$ divides the coefficients of the polynomial

$$P(T, X) - \prod_{i=1}^{n} (X - A_i(T, Y)),$$

for these coefficients vanish at y and Q is the minimal polynomial of y. Therefore, there is a polynomial $R \in \mathbf{Q}(T)[X, Y]$ such that

$$P(T, X) = \prod_{i=1}^{n} (X - A_i(T, Y)) + R(T, X, Y) Q(T, Y). \tag{5.9.9}$$

Similarly, there exists a polynomial $B \in \mathbf{Q}(T)[X_1, \dots, X_n]$ such that $y = B(T, x_1, \dots, x_n)$ and, again, $Q(T, Y)$ divides the coefficients of the polynomial $Y - B(T, A_1(Y), \dots, A_n(Y))$, hence there is a polynomial $S \in \mathbf{Q}(T)[Y]$ such that

$$Y = B(T, A_1(T, Y), \dots, A_n(T, Y)) + S(T, Y) Q(T, Y). \tag{5.9.10}$$

Finally, the polynomial Q is split in K. We thus can find polynomials $C_i \in \mathbf{Q}(T)[Y]$ satisfying

$$Q(T, X) = \prod_{i=1}^{N} (X - C_i(T, y)).$$

As before, it follows that there is a polynomial $U \in \mathbf{Q}(T)[X, Y]$ such that

$$Q(T, X) = \prod_{i=1}^{N} (X - C_i(T, Y)) + U(T, X, Y) Q(T, Y). \tag{5.9.11}$$

The coefficients of the polynomials $P, Q, A_1, \dots, A_n, B, C_1, \dots, C_N, R, S$ belong to $\mathbf{Q}(T)$. Let Σ denote the set of all $t \in \mathbf{Q}$ such that either t is a pole of one of these coefficients, or such that the discriminant of P or Q vanishes at t. By assumption, for any $t \notin \Sigma$, the polynomials $P(t,X)$ and $Q(t,X)$ are separable and the preceding relations hold when evaluated at $T = t$.

Let $t \in \mathbf{Q} \setminus \Sigma$. To prove the lemma, it now suffices to show that the polynomial $P(t,X)$ is split in any extension where $Q(t,X)$ is split, and conversely.

Thus let L be an a extension of $\mathbf{Q}$ in which $Q(t,X)$ has a root η. For any $i \in \{1,\dots,n\}$, let us set $\xi_i = A_i(t,\eta)$. Relation (5.9.9) shows that $P(t,X) = \prod_{i=1}^{n}(X - \xi_i)$, hence $P(t,X)$ is split in L.

Conversely, let L be any extension of $\mathbf{Q}$ in which $P(t,X)$ is split. Denote its roots by $\xi_1,\dots,\xi_n$. Let η be a root of $Q(t,X)$ in some extension L' of L. The roots of P in L' are then given by the $A_i(t,\eta)$, for $1 \leqslant i \leqslant n$, so that there is a permutation $\sigma \in \mathfrak{S}_n$ with $A_i(t,\eta) = \xi_{\sigma(i)}$ for all i. The relation (5.9.10) implies that

$$\eta = B(t, \xi_{\sigma(1)}, \dots, \xi_{\sigma(n)}).$$

It follows that $\eta \in L$ and that $Q(t,X)$ has a root in L.

Now, relation (5.9.11) implies that $Q(t,X) = \prod_{i=1}^{N}(X - C_i(t,\eta))$ is split in L. $\square$

Exercises

Exercise 5.1. **a)** Let G be a finite group and let H be a subgroup of G such that $(G : H) = 2$. Show that H is normal in G.

b) How does this relate to Lemma 5.1.3?

c) More generally, if $(G : H)$ is equal to the smallest prime number dividing $\operatorname{card} G$, show that H is normal in G.

Exercise 5.2. Let $K \subset E$ and $K \subset F$ be two finite extensions with coprime degrees, contained in a common extension Ω of K. Show that $E \cap F = K$ and that $[EF : K] = [E : K][F : K]$.

Exercise 5.3. Let α and β be two distinct complex roots of the polynomial $X^3 - 2$. Let $E = \mathbf{Q}(\alpha)$, $F = \mathbf{Q}(\beta)$.

a) Show that the composite extension $\mathbf{Q} \subset EF$ is a splitting extension of the polynomial $X^3 - 2$ over $\mathbf{Q}$.

b) Show that $E \cap F = \mathbf{Q}$, although $[EF : \mathbf{Q}] \neq [F : \mathbf{Q}]\,[E : \mathbf{Q}]$. (This shows that one cannot remove the hypothesis that one of the extensions E or F is Galois in Corollary 5.3.3.)

Exercise 5.4. This is a sequel to Exercise 1.13, where we showed that the two real roots of the polynomial $P = X^4 - X - 1$ cannot be both constructible with ruler and compass.

a) Show that in fact no root of P is constructible with ruler and compass.

b) What is the Galois group of the extension generated by the complex roots of P?

Exercise 5.5. Let p be a prime number and let $P \in \mathbf{Q}[X]$ be any irreducible polynomial of degree p which has 2 conjugate complex roots, x_1, x_2, and $p-2$ real roots, $x_3, \dots, x_p$. Let us denote by $K = \mathbf{Q}(x_1, \dots, x_p)$ the subfield of $\mathbf{C}$ generated by the roots of P. We identify $\mathrm{Gal}(K/\mathbf{Q})$ with a subgroup of $\mathfrak{S}_p$.

a) Show that the transposition $\tau = (1, 2)$ belongs to $\mathrm{Gal}(K/\mathbf{Q})$. (Think about the complex conjugation.)

b) Show that $\mathrm{Gal}(K/\mathbf{Q})$ contains a p-cycle σ.

c) Show that σ and τ generate $\mathfrak{S}_p$. Conclude that $\mathrm{Gal}(K/\mathbf{Q}) = \mathfrak{S}_p$.

d) *Application*: $P = X^5 - 6X + 3$. (To prove that P is irreducible, use Exercise 1.10 or reduce mod 5.)

Exercise 5.6 (Artin-Schreier's theory). Let p be a prime number. Let K be a field of characteristic p and let $a \in K$. We assume that the polynomial $P = X^p - X - a$ has no root in K. Let $K \subset L$ be any splitting extension of P.

a) If x is a root of P in L, show that the roots of P are x, $x+1$, $x+2$, $\dots$, $x+p-1$. In particular, P is separable.

b) Show that P is irreducible in $K[X]$. (If a degree d polynomial Q divides P, look at the term of degree $d-1$ in Q.)

c) (Another proof that P is irreducible.) Let $x + u$ (for $1 \leqslant u < p$) be another root of the minimal polynomial of x over K. Show that there is $\tau \in \mathrm{Gal}(L/K)$ with $\tau(x) = x + u$. Deduce from this that there is some $\sigma \in \mathrm{Gal}(L/K)$ such that $\sigma(x) = x + 1$, hence that all roots of P are conjugates of x. Conclude.

d) Show that $L = K[x]$ and that $\mathrm{Gal}(L/K) \simeq \mathbf{Z}/p\mathbf{Z}$.

Exercise 5.7 (Cyclic extensions of degree p in characteristic p). Let K be a field of characteristic $p > 0$, and let $K \subset L$ be a finite Galois extension with Galois group $\mathbf{Z}/p\mathbf{Z}$. Let σ be a generator of $\mathrm{Gal}(L/K)$.

a) Show the existence of $t \in L$ such that $\sum_{i=0}^{p-1} \sigma^i(t) = 1$.

Then, set $x = \sum_{i=0}^{p-1} i\sigma^i(t)$.

b) Compute $\sigma(x)$. Show that $x \notin K$ but that $a = x^p - x$ belongs to K.

c) Show that $L = K[x]$ and that $X^p - X - a$ is the minimal polynomial of x over K.

Exercise 5.8. In this exercise, we will determine the Galois group over $\mathbf{Q}$ of the polynomial $P = X^7 - X - 1$, using reduction modulo primes.

a) Show that P has no root in the finite field $\mathbf{F}_8$. Deduce that it is irreducible, when viewed as a polynomial over $\mathbf{F}_2$.

b) Show that the only roots of P in $\mathbf{F}_9$ are the roots of the polynomial X^2+X-1, and that they are simple. Conclude that over $\mathbf{F}_3$, P splits as the product of two irreducible polynomials of degrees 2 and 5.

c) Show that the Galois group of P over the field of rational numbers contains a 7-cycle and a transposition, hence that it is isomorphic to the symmetric group $\mathfrak{S}_7$.

Remark. In fact, for any integer n, the Galois group of the polynomial $X^n - X - 1$ over $\mathbf{Q}$ is equal to $\mathfrak{S}_n$. You may try to prove this by analogous methods for small values of n. If you find the computations too hard, do not hesitate to rely on computer algebra systems, for they often provide routines to factor polynomials modulo prime numbers. For example, the answer to the first question is obtained in less than 1 ms by entering `factormod(x^7-x-1,2)` in PARI/GP, or `Factor(x^7-x-1) mod 2` in MAPLE.

Exercise 5.9 (Another proof of Theorem 5.4.2). Let $K \subset L$ be a finite extension of degree $n \geqslant 2$. Assume that it is Galois and that its Galois group is generated by $\sigma \in \operatorname{Gal}(L/K)$. Assume moreover that $\operatorname{card} \mu_n(K) = n$.

a) Show that $\sigma\colon L \to L$ is a morphism of K-vector spaces, and that its eigenvalues are nth roots of unity.

b) Show that L is the direct sum of the eigenspaces $L_\zeta = \{x \in L\,;\, \sigma(x) = \zeta x\}$, for $\zeta \in \mu_n(K)$.

c) If $y \in L_\zeta \setminus \{0\}$, show that the map $x \mapsto x/y$ is an injective K-linear map $L_\zeta \to L_1$.

d) Show that $L_1 = K$ and conclude that $\dim L_\zeta = 1$ for any $\zeta \in \mu_n(K)$. In particular, if ζ is any primitive nth root of unity, there is a nonzero element $x \in L^*$ such that $\sigma(x) = \zeta x$.

Exercise 5.10. Let $K \subset E$ be a splitting extension of an irreducible separable polynomial $P \in K[X]$. Assume that P has degree n and let $x_1, \dots, x_n$ denote the roots of P in E. One assumes moreover that $\operatorname{Gal}(E/K)$ is cyclic; let σ be a generator.

a) Show that $[E : K] = n$.

b) Assume that $\operatorname{card} \mu_n(K) = n$. For any nth root of unity $\zeta \in K$, define a Lagrange's resolvent by

$$R(\zeta) = x_1 + \zeta\sigma(x_1) + \dots + \zeta^{n-1}\sigma^{(n-1)}(x_1).$$

Show that $R(1) \in K$. For any $\zeta \in \mu_n(K)$, show that $R(\zeta)^n \in K$.

c) Show that E is generated by the $R(\zeta)$ for $\zeta \in \mu_n(K)$.

d) If n is a prime number, show that there is $j \in \{1, \dots, n-1\}$ such that $E = K(\sqrt[n]{R(\zeta)^n})$.

Exercise 5.11. Let K be a field and consider a polynomial $P = X^n - a$, for some $a \in K^*$. Assume that n is not a multiple of the characteristic of K and observe that P is separable.

a) Let L be a splitting extension of K. Show that L contains a primitive nth root of unity ζ. Let $K_1 = K(\zeta)$ and write $\mu_n = \mu_n(K_1)$.

If $m \in \mathbf{Z}$ is prime to n, show that the map $u \mapsto u^m$ is an automorphism of μ_n. Show conversely that any automorphism of μ_n is of this form. Conclude that there is an isomorphism $(\mathbf{Z}/n\mathbf{Z})^* \simeq \operatorname{Aut}(\mu_n)$.

b) Show that the extensions $K \subset K_1$ and $K_1 \subset L$ are Galois, and that their Galois groups are naturally subgroups $A \subset (\mathbf{Z}/n\mathbf{Z})^*$ and $B \subset \mu_n$. (Fix $x \in L$ with $x^n = a$ and look at the action of $\operatorname{Gal}(L/K)$ on x and ζ.)

c) Show that the isomorphism of Question b) restricts to a morphism $\varphi\colon A \to \operatorname{Aut}(B)$ and prove that $\operatorname{Gal}(L/K)$ is isomorphic to the semi-direct product $A \rtimes_\varphi B$.

d) Assume that $[K_1 : K]$ is prime to n and that P is irreducible over K. Show that P is still irreducible over K_1 and that $B = \mu_n$.

e) *Numerical application:* $K = \mathbf{Q}$ and $P = X^7 - 2$. Show that $\operatorname{Gal}(L/K)$ has order 42 and is isomorphic to the group of permutations of $\mathbf{Z}/7\mathbf{Z}$ of the form $n \mapsto an + b$ for $a \in (\mathbf{Z}/7\mathbf{Z})^*$ and $b \in \mathbf{Z}/7\mathbf{Z}$.

Exercise 5.12. This exercise proposes a Galois-theoretic proof of the fundamental theorem of algebra.

Let $\mathbf{R} \subset K$ be a Galois extension of the field of real numbers containing the field of complex numbers $\mathbf{C}$. Let $G = \operatorname{Gal}(K/\mathbf{R})$ and let P be a 2-Sylow subgroup of G. Set $\operatorname{card} P = 2^n$.

a) Using the fact that $\mathbf{R}$ has no finite extension of odd degree, show that $G = P$.

b) Let $P_1 = \operatorname{Gal}(K/\mathbf{C})$. By Lemma 5.1.3, P has a normal series

$$\{1\} = P_n \subset \cdots \subset P_2 \subset P_1 \subset P$$

with $(P_{j+1} : P_j) = 2$ for any j. Define $K_j = K^{P_j}$. Show that the extension $K_j \subset K_{j+1}$ is a quadratic extension. Using the fact that any complex number is a square, show that $n = 1$, hence $K = \mathbf{C}$.

Exercise 5.13. This exercise will let you prove Theorem 5.1.1 without any group theory, using instead ideas from the second proof of the fundamental theorem of algebra.

Let z be any algebraic number, and assume that the degree of the extension of $\mathbf{Q}$ generated by its conjugates $z_1, \ldots, z_d$ is a power of 2.

Observe that d is itself a power of 2. By induction on d, prove as follows that z is constructible.

a) Fix $c \in \mathbf{Q}$, set $z_{i,j,c} = z_i + z_j + cz_iz_j$ and $Q_c = \prod_{i<j} (X - z_{i,j,c})$. Show that Q_c is a polynomial with rational coefficients, and that the degrees of its irreducible factors are powers of 2. Show that at least one of these degrees divides $d/2$, hence that there are $i < j$ such that $z_i + z_j + cz_iz_j$ is constructible.

b) Show that there are i and j such that $z_i + z_j$ and z_iz_j are constructible. Conclude that z_i and z_j are both constructible.

c) Show that z is constructible.

Exercise 5.14. Let n be an integer, with $n \geqslant 5$. Let $K \subset L$ be a finite Galois extension with Galois group $\mathfrak{S}_n$.

a) Show that there is only one quadratic extension $K \subset K_1$ contained in L. What is the Galois group of the extension $K_1 \subset L$? (Use Exercise 4.17.)

b) Show that the degree of any $x \in L \setminus K_1$ is at least n.

Exercise 5.15. Let K be a field, and let $\varphi \colon K \to k \cup \{\infty\}$ be a place of K. Recall that we defined the valuation ring of φ as the set $A = \{x \in K \,;\, \varphi(x) \neq \infty\}$.

a) Show also that for any $x \in K \setminus \{0\}$, either x or $1/x$ belongs to A (this is the general definition of a valuation ring).

b) Let $\mathfrak{m} = \varphi^{-1}(0)$. Show that $\mathfrak{m}$ is an ideal of A and that an element $a \in A$ is invertible in A if and only if $a \notin \mathfrak{m}$.

c) Deduce from this that $\mathfrak{m}$ is the unique maximal ideal in A, that $A/\mathfrak{m}$ is a field, and that φ induces a field homomorphism $A/\mathfrak{m} \to k$.

d) In the two examples given in the text (Example 5.8.2), show that the ideal $\mathfrak{m}$ is generated by one element π. Show moreover that any ideal in A is generated by a power of π. (In fact, one can set $\pi = p$ in case *a*) and $\pi = X - \alpha$ in case *b*).) In particular, in these two cases, the ring A is a principal ideal ring.

Exercise 5.16. Let K be a field and let A be a subring in K. Fix an algebraic closure Ω of K. One says that an element $x \in \Omega$ is *integral* over A if there is a monic polynomial $P \in A[X]$ such that $P(x) = 0$.

a) Let x and y be two elements in Ω which are integral over A. Let P and $Q \in A[X]$ be monic polynomials such that $P(x) = Q(y) = 0$. Factor P and Q in Ω as

$$P = \prod_{i=1}^{n}(X - x_i) \quad \text{and} \quad Q = \prod_{j=1}^{m}(X - y_j).$$

Show that the coefficients of the polynomial $R = \prod_{i,j}(X - x_i - y_j)$ belong to A. (Write $R = \prod_i Q(X - x_i)$ and use the theorem on symmetric polynomials.) Conclude that $x + y$ is integral over A. Similarly, show that xy is integral over A.

b) Show that the set of elements of Ω which are integral over A form a subring of Ω.

c) Assume that A is a valuation ring. Show that an element $x \in K$ is integral over A if and only if $x \in A$. ("A valuation ring is integrally closed.")

d) Let P and Q be two monic polynomials in $K[X]$. Assume that $P \in A[X]$ and that Q divides P in $K[X]$. Show that the coefficients of Q are integral over A.

e) Assuming that A is a valuation ring, conclude that $Q \in A[X]$. ("Gauss's lemma for valuation rings.")

6

Algebraic theory of differential equations

In this final chapter, I want to explain how certain aspects of the theory of linear differential equations with, say, polynomial coefficients, can be viewed in an algebraic setting. There is in fact a full "Galois theory of differential equations" of which I try to convey some ideas. I conclude with a theorem due to Liouville, a particular case of which is the fact that the function $\int \exp(x^2)\,dx$ *has no elementary algebraic expression.*

6.1 Differential fields

Definition 6.1.1. *Let A be a ring.* A derivation *on A is a homomorphism of abelian groups $D\colon A \to A$ which satisfies the* Leibniz rule*: for any a and b in A, one has*

$$D(ab) = aD(b) + bD(a).$$

A differential ring *is a ring endowed with a derivation. When the ring is a field, we call it a* differential field*. One often denotes $a' = D(a)$, $a'' = D(D(a))$, and, for any integer $n \geqslant 0$, $a^{(n)} = D^n(a)$.*

Examples 6.1.2 (Examples of differential rings).

a) The ring of $\mathscr{C}^\infty$ functions on an interval $I \subset \mathbf{R}$ with, say, complex values, endowed with the usual derivation of functions, i.e. , we set $D(f)$ to be the derivative of f.

b) If X is a manifold, a derivation on the ring of smooth functions on X is also called a *vector field* on X.

c) The ring of real analytic functions on an open interval in $\mathbf{R}$, endowed with the usual derivation.

d) The ring of holomorphic functions on an open subset of $\mathbf{C}$, endowed with the derivation $f \mapsto f'$.

e) The ring $k[T]$ of polynomials on one variable T, with coefficients in a field k, together with the formal derivation $P \mapsto P'$.

f) Any ring A, with the identically zero derivation defined by $D(a) = 0$ for any $a \in A$ (stupid example).

Examples 6.1.3 (Examples of differential fields).

a) The field $k(T)$ of rational functions in one variable with coefficients in a field k, endowed with the formal derivation of rational functions.

b) The field of meromorphic functions on a connected open subset of $\mathbf{C}$, with the usual derivation.

Differential rings or fields feature the following familiar formulae.

Lemma 6.1.4. *Let (A, D) be a differential ring. Let a, b be two elements of A.*

a) $D(1) = 0$;

b) *for any integer* $n \geqslant 1$, $D(a^n) = na^{n-1}D(a)$;

c) *for any integer* $n \geqslant 1$, $D^n(ab) = \sum\limits_{k=0}^{n} \binom{n}{k} D^k(a)D^{n-k}(b)$;

d) *if b is invertible, then* $D(a/b) = (bD(a) - aD(b))/b^2$. *In particular,* $D(1/b) = -D(b)/b^2$.

Proof. *a*) Applying the derivation D to both sides of the equality $1 \times 1 = 1$, one gets $1D(1) + 1D(1) = D(1)$, hence $D(1) = 0$. More generally, one has $D(n \cdot 1) = 0$ for any $n \in \mathbf{Z}$.

b) Let us prove this by induction on n. The formula is true for $n = 1$. If it holds for n, then

$$
\begin{aligned}
D(a^{n+1}) &= D(a \times a^n) = aD(a^n) + a^n D(a) \\
&= a(na^{n-1}D(a)) + a^n D(a), = (n+1)a^n D(a)
\end{aligned}
$$

hence it holds for $n + 1$.

c) Let us again prove this formula by induction on n. It holds for $n = 1$. Assuming it holds for n, then

$$
\begin{aligned}
D^{n+1}(ab) = D\big(D^n(ab)\big) &= D\left(\sum_{k=0}^{n} \binom{n}{k} D^k(a)D^{n-k}(b)\right) \\
&= \sum_{k=0}^{n} \binom{n}{k} D\left(D^k(a)D^{n-k}(b)\right) \\
&= \sum_{k=0}^{n} \binom{n}{k} \left(D^{k+1}(a)D^{n-k}(b) + D^k(a)D^{n+1-k}(b)\right)
\end{aligned}
$$

$$= \sum_{k=1}^{n+1} \binom{n}{k-1} D^k(a) D^{n+1-k}(b) + \sum_{k=0}^{n} \binom{n}{k} D^k(a) D^{n+1-k}(b)$$
$$= \sum_{k=0}^{n+1} \binom{n+1}{k} D^k(a) D^{n+1-k}(b)$$

by virtue of the classical formula

$$\binom{n+1}{k} = \binom{n}{k-1} + \binom{n}{k},$$

valid for $n \geqslant 1$ and $k \geqslant 1$.

d) Differentiating the relation $b(a/b) = a$, one gets

$$D(b)\frac{a}{b} + bD(a/b) = D(a),$$

hence

$$D(a/b) = \frac{D(a)}{b} - D(b)\frac{a}{b^2} = \frac{bD(a) - aD(b)}{b^2},$$

as was to be shown. The last relation follows since $D(1) = 0$. □

An element of a differential ring is said to be *constant* if its derivative is zero.

Proposition 6.1.5. *The set of constant elements in a differential ring is a subring. The set of constant elements in a differential field is a subfield, called the* constant field.

Proof. If a and b are elements of a differential ring (A, D) satisfying $D(a) = D(b) = 0$, one has $D(a+b) = D(a)+D(b) = 0$ and $D(ab) = aD(b)+bD(a) = 0$. Since $D(1) = 0$, the set of all $a \in A$ with $D(a) = 0$ is a subring of A.

If $a \in A$ is both constant and invertible, the preceding lemma shows that $D(1/a) = 0$, hence $1/a$ is constant. In particular, if (K, D) is a differential field, the set K^D of all $x \in K$ with $D(x) = 0$ is a subfield of K. □

One often denotes A^D, *resp.* K^D the set of constant elements in a differential ring (A, D), *resp.* in a differential field (K, D). In all examples above coming from analysis, the constant elements are the (locally) constant functions. For polynomials in characteristic p, something funny happens.

Proposition 6.1.6. *Let k be a field. Set $A = k[T]$ and $K = k(T)$, endowed with the usual derivation. If k has characteristic 0, $A^D = K^D = k$. If k has characteristic $p > 0$, then $A^D = k[T^p]$ and $K^D = k(T^p)$.*

Proof. Let $P = \sum_{n=0}^{N} a_n T^n$, then $P' = \sum_{n=0}^{N} n a_n T^{n-1}$. Assume that $P' = 0$, hence $na_n = 0$ for any integer n. If k has characteristic zero, this implies $P = a_0$. However, if the characterstic of k is $p > 0$, this only implies that $a_n = 0$ whenever p does not divide n, hence $P \in K[T^p]$. The other inclusion is obvious.

Now look at rational functions and let $R \in k(T)$ be such that $R' = 0$. Write $R = A/B$ where A and B are two polynomials, the polynomial B being $\neq 0$ and of minimal degree. It follows that $BR = A$, and by differentiating both sides, we obtain $B'R = A'$. The degree of B' is smaller than the degree of B, so the minimality assumption implies that $B' = 0$, hence $A' = 0$. If k has characteristic 0, it follows that A and B are constant, hence $k(T)^D = k$. If k has characteristic $p > 0$, A and B are two polynomials in the variable T^p, so $R \in k(T^p)$. Conversely, such rational functions have a zero derivative, q.e.d.□

6.2 Differential extensions; construction of derivations

Definition 6.2.1. *A differential homomorphism $f\colon (A, D_A) \to (B, D_B)$ from one differential ring to another is a ring homomorphism $f\colon A \to B$ such that for any $a \in A$, $f(D_A(a)) = D_B(f(a))$.*

If A and B are fields, one speaks of differential extension of fields, *or simply of* differential extension.

If no confusion about the morphism $f\colon A \to B$ can arise, one also says that D_B *extends* D_A.

Lemma 6.2.2. *Let $f\colon (A, D_A) \to (B, D_B)$ be a differential homomorphism of rings. The kernel of f is stable under D_A.*

Indeed, for any $a \in A$ with $f(a) = 0$, one has $f(D_A(a)) = D_B(f(a)) = D_B(0) = 0$. One says that $\operatorname{Ker} f$ is a *differential ideal.*

Conversely, let I be a differential ideal in (A, D_A) and let us show how the quotient ring $B = A/I$ can be endowed with a canonical structure of differential ring, such that the ring morphism $\pi\colon A \to B$ is a differential homomorphism. By definition, the morphism of abelian groups

$$\pi \circ D_A\colon A \to B$$

is zero on I. Since I is a subgroup of A and since A is abelian, there is a unique morphism of abelian groups $D_B\colon B \to B$ such that $D_B(\pi(x)) = \pi(D_A(x))$ for any $x \in A$. Let us now show that D_B is a derivation. Let a and b be two elements of B; let x and y in A be such that $a = \pi(x)$ and $b = \pi(y)$. Then, one has

$$\begin{aligned}
D_B(ab) &= D_B(\pi(x)\pi(y)) = D_B(\pi(xy)) \\
&= \pi(D_A(xy)) && \text{by definition of } D_B \\
&= \pi(yD_A(x) + xD_A(y)) && \text{since } D_A \text{ is a derivation} \\
&= \pi(y)\pi(D_A(x)) + \pi(x)\pi(D_A(y)) \\
& \qquad\qquad \text{since } \pi \text{ is a ring homomorphism} \\
&= \pi(y)D_B(\pi(x)) + \pi(x)D_B(\pi(y)) \\
&= bD_B(a) + aD_B(b).
\end{aligned}$$

Theorem 6.2.3. *Let (A, D_A) be a ring and let I be a differential ideal of A. Then there exists a unique derivation of the quotient ring A/I such that the canonical ring morphism $A \to A/I$ is a differential homomorphism.*

Proposition 6.2.4. *Let (A, D) be a differential ring. Assume that A is an integral domain and let K be its field of fractions. There exists a unique derivation on K which coincides with D on A.*

Consequently, K has a canonical structure of a differential field.

Proof. Keeping in mind the formulae of Section 6.1, one necessarily has to set, if $x \in K$ is the quotient a/b of two elements of A,

$$D(x) = \frac{D(a)b - aD(b)}{b^2}.$$

Let us check that this formula does not depend on the choice of the fraction a/b and that it defines a derivation on K. For any $t \in A \setminus \{0\}$, one has

$$\begin{aligned}
\frac{D(at)(bt) - (at)D(bt)}{(bt)^2} &= \frac{D(a)bt^2 + abtD(t) - at^2D(b) - atbD(t)}{b^2t^2} \\
&= \frac{D(a)b - D(b)a}{b^2}.
\end{aligned}$$

Consequently, the formulae for $D(a/b)$ and $D(ad/bd)$ give the same result, and similarly, the formulae for $D(c/d)$ and $D(bc/bd)$ give the same result. Since $ad = bc$, the formulae for $D(a/b)$ and $D(c/d)$ compute the same element of K, hence the map $D\colon K \to K$ is well-defined.

Moreover, if $x = a/b$ and $y = c/d$, one has

$$\begin{aligned}
D(x+y) &= D\Big(\frac{ad+bc}{bd}\Big) = \frac{D(ad+bc)bd - (ad+bc)D(bd)}{b^2d^2} \\
&= \frac{D(ad)bd - adD(bd)}{b^2d^2} + \frac{D(bc)bd - bcD(bd)}{b^2d^2} \\
&= D\Big(\frac{ad}{bd}\Big) + D\Big(\frac{bc}{bd}\Big) = D(a/b) + D(c/d) = D(x) + D(y).
\end{aligned}$$

It follows that D is a homomorphism of abelian groups. On the other hand,

$$\begin{aligned} D(xy) = D\Big(\frac{ac}{bd}\Big) &= \frac{D(ac)bd - acD(bd)}{b^2d^2} \\ &= \frac{abdD(c) + bcdD(a) - acdD(b) - abcD(d)}{b^2d^2} \\ &= \frac{bcdD(a) - acdD(b)}{b^2d^2} + \frac{abdD(c) - abcD(d)}{b^2d^2} \\ &= \frac{bD(a) - aD(b)}{b^2}\frac{cd}{d^2} + \frac{dD(c) - cD(d)}{d^2}\frac{ab}{b^2} \\ &= D\Big(\frac{a}{b}\Big)\frac{c}{d} + D\Big(\frac{c}{d}\Big)\frac{a}{b} = D(x)y + D(y)x, \end{aligned}$$

which shows that D is a derivation. □

I now explain how to construct all the derivations on a polynomial ring.

Theorem 6.2.5. *Let (A, D) be a differential ring and consider the ring $A[T]$ of polynomials in one variable T with coefficients in A. For any $b \in A[T]$, there is a unique derivation D_b of $A[T]$ with $D_b(T) = b$ such that the canonical ring morphism $A \to A[T]$ is a differential morphism $(A, D) \to (A[T], D_b)$.*

Proof. Denote $B = A[T]$ and let $P = \sum_{k=0}^{n} a_k T^k$ be an element of B. If D_B is any derivation of B extending D, one has

$$\begin{aligned} D_B(P) &= \sum_{k=0}^{n} D_B(a_k T^k) = \sum_{k=0}^{n} \big(D(a_k)T^k + a_k D_B(T^k)\big) \\ &= \sum_{k=0}^{n} D(a_k)T^k + \Big(\sum_{k=0}^{n} k a_k T^{k-1}\Big) D_B(T) \\ &= P^D(T) + D_B(T)P'(T), \end{aligned}$$

where P^D denotes the polynomial of $A[T]$ obtained by applying D to the coefficients of P. This formula shows that such a derivation is determined by the image $D_B(T)$ of T. Conversely, let $\lambda \in B$ and let us show that the formula

$$D_B(P) = P^D(T) + \lambda P'(T) = \sum_{k=0}^{n} D(a_k)T^k + \lambda \sum_{k=0}^{n} k a_k T^{k-1}$$

defines a derivation on B satisfying $D_B(T) = \lambda$ and extending the derivation D on A. The map D_B is obviously a morphism of abelian groups. If $Q = \sum_{k=0}^{m} b_k T^k$ is another polynomial, one has

$$PQ = \sum_{k=0}^{m+n} c_k T^k, \qquad c_k = \sum_{i+j=k} a_i b_j$$

and

$$
\begin{aligned}
D_B(PQ) &= \sum_{k=0}^{m+n} D(c_k)T^k + \lambda \sum_{k=0}^{n} kc_k T^{k-1} \\
&= \sum_{k=0}^{m+n} \sum_{i+j=k} \left(D(a_i)b_j + D(b_j)a_i\right) T^{i+j} \\
&\qquad\qquad + \lambda \sum_{k=0}^{n} \sum_{i+j=k} (i+j)a_i b_j T^{i+j-1} \\
&= \sum_{i=0}^{n} \sum_{j=0}^{m} D(a_i)b_j T^{i+j} + \lambda \sum_{i=0}^{n} \sum_{j=0}^{m} i a_i b_j T^{i+j-1} \\
&\qquad + \sum_{i=0}^{n} \sum_{j=0}^{m} D(b_j)a_i T^{i+j} + \lambda \sum_{i=0}^{n} \sum_{j=0}^{n} j a_i b_j T^{i+j-1} \\
&= D_B(P)Q + D_B(Q)P,
\end{aligned}
$$

which was to be proved. □

A last case, very important in the following discussion, concerns (separable) algebraic extensions.

Theorem 6.2.6. *Let (K, D) be a differential field and let L be a finite separable algebraic extension of K. Then there exists a unique derivation on L which extends D.*

Proof. The proof is nothing but an abstract algebraic version of the computation of the derivative of a function defined implicitly.

Let $x \in L$ be any primitive element, so that $L = K[x]$; denote by $P = X^n + a_{n-1}X^{n-1} + \cdots + a_0$ its minimal polynomial, hence $L \simeq K[X]/(P)$. If D_L is a derivation of L which extends that of K, one obtains by differentiating the relation $P(x) = 0$ that

$$
\begin{aligned}
0 = D_L(0) &= D_L(P(x)) \\
&= \sum_{k=0}^{n} D(a_k)x^k + \sum_{k=0}^{n} k a_k x^{k-1} D_L(x) \\
&= P^D(x) + P'(x)D_L(x).
\end{aligned}
$$

(We denoted by P^D the polynomial obtained by applying D to the coefficients of P.) One has $\deg P' < \deg P$ and, since P is separable, $P' \neq 0$. Hence $P'(x) \neq 0$, because P is the minimal polynomial of x. Consequently,

$$
D_L(x) = -P^D(x)/P'(x)
$$

and there can be at most one derivation on L extending the given derivation on K. To show that such a derivation actually exists, we will use Theorem 6.2.3. We need to show that there is a derivation on $K[X]$ such that the ideal (P) is a differential ideal. If $\tilde{D}$ is a derivation of $K[X]$, the preceding computation shows that

$$\tilde{D}(P) = P^D + P'(X)\tilde{D}(X).$$

Since P is separable, P and P' are coprime and there exist polynomials U and $V \in K[X]$ such that $UP + VP' = 1$. Then the choice $\tilde{D}(X) = -VP^D$ defines a derivation $\tilde{D}$ on $K[X]$ such that

$$\tilde{D}(P) = P^D - VP^DP' = (1 - VP')P^D = (UP^D)P.$$

This is a multiple of P. Consequently, for any $A \in K[X]$, $\tilde{D}(AP) = \tilde{D}(A)P + A\tilde{D}(P) \in (P)$ and the ideal (P) is a differential ideal of the differential ring $(K[X], \tilde{D})$. The quotient ring $L = K[X]/(P)$ inherits the desired structure of a differential field. □

6.3 Differential equations

Let (K, D) be a differential field. The differential equations we are interested in have the form

$$D^n(f) + a_{n-1}D^{n-1}(f) + \cdots + a_0 f = 0,$$

where $a_0, \ldots, a_{n-1} \in K$, the unknown being f. In other words, we will only discuss *linear homogeneous* differential equations. As in calculus, we will say that the preceding differential equation has order n. Actually, we will rather consider differential equations in matrix form

$$Y' = AY, \quad A \in M_n(K),$$

the unknown being a vector Y, written as a column (the derivative of such a vector is defined by differentiating each coordinate).

As in calculus again, one can turn an equation of the first sort into an equation of the second one: just introduce the vector $Y = (f, f', \ldots, f^{(n-1)})^{\mathrm{t}}$. One then has

$$\begin{aligned} Y' &= (f', f'', \ldots, f^{(n)})^{\mathrm{t}} \\ &= (f', f'', \ldots, f^{(n-1)}, -a_{n-1}f^{(n-1)} - \cdots - a_0 f)^{\mathrm{t}} \\ &= \begin{pmatrix} 0 & 1 & 0 & \\ & \ddots & \ddots & \\ & & 0 & 1 \\ -a_0 & -a_1 & \ldots & -a_{n-1} \end{pmatrix} Y. \end{aligned}$$

It could be possible to consider vector-valued differential equations of higher order. They can be reduced to a first-order differential equation by a similar procedure.

Theorem 6.3.1. *Let (K, D) be a differential field and let C denote its field of constants. Then the set of solutions $Y \in K^n$ of a differential equation $Y' = AY$, with $A \in M_n(K)$, is a C-vector space of dimension less than or equal to n.*

Proof. Observe that the derivation $D\colon K \to K$ is C-linear (for $a \in C$ and $f \in K$, $D(af) = aD(f) + D(a)f = aD(f)$), so that the map $\varphi\colon K^n \to K^n$ defined by $\varphi(Y) = Y' - AY$ is C-linear. Its kernel, the set of solutions of the differential equation $Y' = AY$, is therefore a C-vector space, which we denote by V.

Let us show that its dimension is $\leqslant n$. It suffices to show that $n+1$ elements in V, say $Y_0, \dots, Y_n$, are linearly dependent over C. They are obviously dependent over K, since the dimension of K^n as a K-vector space is n. Hence we are reduced to proving the following lemma. □

Lemma 6.3.2. *Let (K, D) be a differential field with field of constants C. Let $Y_1, \dots, Y_m$ be solutions of a differential equation $Y' = AY$, for $A \in M_n(K)$. If they are linearly independent over C, then they are linearly independent over K.*

Proof. Let us show this by induction on m. For $m = 1$, the hypothesis and the conclusion both mean that $Y_1 \neq 0$. Assume that the result holds for $(m-1)$. By induction, we may assume that $Y_1, \dots, Y_{m-1}$ are linearly independent over K. Let us consider a linear relation $a_1 Y_1 + \dots + a_m Y_m = 0$, for $a_1, \dots, a_m \in K$. Necessarily, $a_m \neq 0$, which allows us to divide this relation by a_m, hence we assume $a_m = 1$. Now, let us differentiate this relation, obtaining

$$(a_1' Y_1 + a_1 Y_1') + \dots + (a_{m-1}' Y_{m-1} + a_{m-1} Y_{m-1}') + Y_m' = 0,$$

that is,

$$(a_1' Y_1 + \dots + a_{m-1}' Y_{m-1}) + A(a_1 Y_1 + \dots + a_{m-1} Y_{m-1} + Y_m) = 0,$$

hence

$$a_1' Y_1 + \dots + a_{m-1}' Y_{m-1} = 0.$$

This is a linear dependence relation with coefficients in K for $Y_1, \dots, Y_{m-1}$. By hypothesis, it is trivial and $a_1' = \dots = a_{m-1}' = 0$. In other words, $a_1, \dots, a_{m-1}$ are constants and $Y_1, \dots, Y_m$ are linearly dependent over C, a contradiction. □

There is a nice tool in linear algebra to detect the linear independence over the field of constants C, given by the Wronskian construction.

Definition 6.3.3. *Let (K, D) a differential field. The* Wronskian *of n elements $f_1, \dots, f_n \in K$ is defined as the determinant*

$$W(f_1, \dots, f_n) = \det \begin{pmatrix} f_1 & f_2 & \dots & f_n \\ f_1' & f_2' & \dots & f_n' \\ \vdots & & & \vdots \\ f_1^{(n-1)} & f_2^{(n-1)} & \dots & f_n^{(n-1)} \end{pmatrix}.$$

Theorem 6.3.4. *Let (K, D) be a differential field. Elements $f_1, \dots, f_n$ in K are linearly dependent over C if and only if their Wronskian is zero.*

Proof. This is a variant of the preceding proof. If $f_1, \dots, f_n$ are elements of K that are linearly dependent over C, one immediately sees that the columns of the Wronskian matrix satisfy a linear relation, hence $W(f_1, \dots, f_n) = 0$. The important point is the converse, which we shall prove by induction on n, the result being clearly true for $n = 1$. Assume now that $W(f_1, \dots, f_n) = 0$. If $W(f_2, \dots, f_n) = 0$, it follows by induction that $f_2, \dots, f_n$ are linearly dependent over C. Assume therefore that $W(f_2, \dots, f_n) \neq 0$. Since $W(f_1, \dots, f_n) = 0$, the columns of the Wronskian matrix satisfy a nontrivial linear dependence relation with coefficients in K, say

$$a_1 f_1^{(j)} + a_2 f_2^{(j)} + \dots + a_n f_n^{(j)} = 0, \qquad 0 \leqslant j \leqslant n - 1. \tag{$*_j$}$$

Since by assumption $W(f_2, \dots, f_n) \neq 0$, one has $a_1 \neq 0$ and we may assume, dividing by a_1, that $a_1 = 1$. Now differentiating the relations $(*_j)$ for $0 \leqslant j \leqslant n-2$, one gets

$$f_1^{(j+1)} + (a_2 f_2^{(j+1)} + a_2' f_2^{(j)}) + \dots + (a_n f_n^{(j+1)} + a_n' f_n^{(j)}) = 0,$$

hence

$$a_2' f_2^{(j)} + \dots + a_n' f_n^{(j)} = 0, \qquad 0 \leqslant j \leqslant n - 2.$$

Were they not trivial, these relations would imply that $W(f_2, \dots, f_n) = 0$. Consequently, $a_2' = \dots = a_n' = 0$ and the a_j are all constants, which shows that $f_1, \dots, f_n$ are linearly dependent over C. □

6.4 Picard-Vessiot extensions

Recall that for any polynomial, we have defined and constructed a splitting extension as a minimal extension where this polynomial has a full set of roots. Similarly, we shall now construct a minimal extension of a differential field in which a nth order differential equation admits n linearly independent solutions.

In the following, we consider only fields of characteristic zero.

Definition 6.4.1. *Let (K, D) be a differential field. Assume that the field C of constants in K is algebraically closed with characteristic zero. Let* (E)*: $Y' = AY$ be a linear homogeneous differential equation, where A is a $n \times n$ matrix with coefficients in K.*

One says that a differential extension (L, D) of K is a Picard-Vessiot extension *for this equation if*

a) *the vector space of solutions of* (E) *in L^n has dimension n, hence a basis of solutions $(Y_1, \ldots, Y_n)$ with coefficients in L;*
b) *L is generated by the coefficients Y_{ij} of this basis;*
c) *the field of constants of L is equal to C.*

Theorem 6.4.2. *Any differential equation admits a Picard-Vessiot extension. Two such extensions are isomorphic as differential extensions of (K, D).*

The proof is quite complicated and we shall only establish the existence of a Picard-Vessiot extension.

Proof that a Picard-Vessiot extension exists. Since a Picard-Vessiot extension is generated by the coefficients of a basis of solutions, let us begin by considering the ring

$$R = K[Y_{11}, \ldots, Y_{nn}]$$

of polynomials in n^2 indeterminates. If G denotes the matrix (Y_{ij}), let us endow the ring R with the derivation $D \colon R \to R$ defined by $D(G) = AG$, which means that we have *added* to K a family of solutions to equation (E). The requirement that the solutions are linearly independent can be rephrased as the fact that $\det(G)$ is invertible. Let us therefore introduce the ring $S = R[T]/(1 - T\det(G))$, in which T corresponds to the inverse of $\det(G)$. We have to extend the derivation D from R to S. To that aim, we have to define $D(T)$ so that the ideal generated by $1 - T\det(G)$ becomes a differential ideal. By Exercice 6.2, the derivative of $\det(G)$ is given by

$$D(\det G) = \operatorname{Tr}(\operatorname{Com}(G)G'),$$

where $\operatorname{Com}(G)$ denotes the comatrix of G (transpose of the matrix of cofactors). Since $D(G) = AG$, we thus have

$$\begin{aligned} D(\det G) &= \operatorname{Tr}(\operatorname{Com}(G)AG) = \operatorname{Tr}(AG\operatorname{Com}(G)) \\ &= \operatorname{Tr}(A\det(G)) = \operatorname{Tr}(A)\det(G). \end{aligned}$$

Consequently, one has

$$\begin{aligned} D(1 - T\det(G)) &= -D(T)\det(G) + T\operatorname{Tr}(A)\det(G) \\ &= -\det(G)(D(T) - T\operatorname{Tr}(A)). \end{aligned}$$

In other words, the ideal $(1 - T\det(G))$ is a differential ideal as soon as the derivative of T satisfies $D(T) = T\operatorname{Tr}(A)$. Define $D(T)$ by this relation. It follows that the quotient ring $S = K[Y_{11}, \dots, Y_{nn}, T]/(1 - T\det(G))$ inherits a derivation such that $D(G) = AG$.

But we have not yet finished: the differential ring S that we just constructed is not a field, and has in general far too many constant elements (see Exercise 6.4). Therefore, let I be a differential ideal of S, maximal among all differential ideals of S distinct from S. (A transfinite induction similar to that of Theorem 2.5.3 shows that there are such ideals.) The quotient ring S/I is a differential ring and it has no nontrivial differential ideal. By Lemma 6.4.3 below, it is an integral domain and the field of constants of its field of fractions is equal to C. This shows that (L, D) is a Picard-Vessiot extension for the equation $Y' = AY$. □

Lemma 6.4.3. *Let (K, D) be a differential field of characteristic zero; let us denote its field of contstants by C. Consider a morphism of differential rings $(K, D) \to (A, D)$ and assume that A has no differential ideal except* (0) *and A. Then the following hold:*

a) *The ring A is integral. Let us denote by L its field of fractions, endowed with its natural derivation.*

b) *The field of constants of L is contained in A.*

c) *If A is a finitely generated K-algebra and if C is algebraically closed, then the field of constants of L is equal to C.*

Proof. a) Let us begin to show that A contains no nonzero nilpotent elements. To that aim, let I be the set of all $x \in I$ a power of which is zero. This is an ideal of A (see Exercise 2.10) and we shall show that it is even a differential ideal. Indeed, let $x \in I$ and let $n \geqslant 1$ be any integer such that $x^n = 0$. Differentiate this relation: one gets $nx^{n-1}x' = 0$, hence $x^{n-1}x' = 0$ since $K \subset A$ has characteristic 0. We will now prove by induction that for any integer k with $0 \leqslant k \leqslant n$, one has $x^{n-k}(x')^{2k} = 0$. This is indeed true for $k = 0$ and $k = 1$. Differentiating the relation $x^{n-k+1}(x')^{2k-2}$ for $k \leqslant n$ gives

$$(n - k + 1)x^{n-k}(x')^{2k-1} + 2(k - 1)x^{n-k+1}(x')^{2k-3}x'' = 0.$$

Multiplying by x', we obtain

$$(n - k + 1)x^{n-k}(x')^{2k} = 0.$$

Since $k \leqslant n$, $n - k + 1 \neq 0$ is invertible in K (using again the hypothesis that K has characteristic 0) and we find that $x^{n-k}(x')^{2k}$. When $k = n$, one gets $(x')^{2n} = 0$, hence $x' \in I$. This completes the proof that I is a differential ideal. Since 1 is not nilpotent, $I \neq A$, and the assumption that A has no nontrivial differential ideal implies that $I = 0$.

Let now a be any nonzero element of A and let us prove that a is not a zero-divisor. Let I be the set of all $b \in B$ such that $ab = 0$. This is an ideal of A. Moreover, if $ab = 0$, then one has $ab' + a'b = 0$, hence $a^2b' = 0$ after multiplying by a. It follows that $(ab')^2 = 0$ and $ab' = 0$ since A has no nonzero nilpotent elements. Consequently $b' \in I$ and I is a differential ideal. Since $a \neq 0$, $1 \notin I$ and $I = 0$. It follows that A is an integral domain.

b) Let us denote by C' the field of constants of L. It is a subfield of L containing C. Let $x \in C'$ and let I be the set of all $a \in A$ such that $ax \in A$. This is an ideal of A. It is even a differential ideal of A. Indeed, if $b = ax \in A$, then $b' = ax' + a'x = a'x \in A$, hence $a' \in I$. By definition of the field of fractions, $I \neq 0$, hence $I = A$. In particular, $1 \in I$ and $x = 1x \in A$.

c) It follows from *b*) that $C' \subset A$. Let $\mathfrak{m}$ be any maximal ideal of A. The quotient ring $A/\mathfrak{m}$ is a field. By assumption, A is a finitely generated K-algebra; let $x_1, \dots, x_n$ be elements of A such that $A = K[x_1, \dots, x_n]$. Then the images in $A/\mathfrak{m}$ of $x_1, \dots, x_n$ generate $A/\mathfrak{m}$ as a K-algebra, so that $A/\mathfrak{m}$ is a finitely generated K-algebra too. By Hilbert's Nullstellensatz (Theorem 6.8.1), $A/\mathfrak{m}$ is an algebraic extension of K. The morphism $C' \to A \to A/\mathfrak{m}$ is a morphism of fields, hence is injective. It follows that C' is algebraic over K. By Lemma 6.4.4, C' is moreover algebraic over the field of constants of K. As C is algebraically closed, the inclusion map $C \to C'$ is an isomorphism and $C' = C$. □

Lemma 6.4.4. *Let $(K, D) \to (L, D)$ a differential homomorphism of differential fields of characteristic zero. Denote by C the field of constants in K. Let $x \in L$. The following are equivalent:*

a) *x is algebraic over C;*
b) *x is constant and algebraic over K.*

Proof. Assume that x is algebraic over C. It is *a fortiori* algebraic over K; we have to show that x is constant. Let $P = X^n + a_{n-1}X^{n-1} + \cdots + a_0$ be the minimal polynomial of x over C. Differentiating the relation $x^n + a_{n-1}x^{n-1} + \cdots + a_0 = 0$, one gets $P'(x)x' = 0$. Since K has characteristic zero, P is separable and $P'(x) \neq 0$. Therefore $x' = 0$.

Assume now that $x \in L$ is constant and algebraic over K. Let $P = X^n + a_{n-1}X^{n-1} + \cdots + a_0$ its minimal polynomial over K. Let us differentiate again the relation $P(x) = 0$. Since $x' = 0$, it follows that $\sum_{k=0}^{n-1} a_k' x^k = 0$, that is, $P^D(x) = 0$. Since $\deg P^D < \deg P$, one has $P^D = 0$ and $a_k' = 0$ for any k. This means that $P \in C[X]$, hence x is algebraic over C. □

6.5 The differential Galois group; examples

Let (K, D) be a differential field. We assume that its field of constants C is an algebraically closed field of characteristic zero. Let (L, D) be a Picard-Vessiot extension of K corresponding to an equation (E): $Y' = AY$. Let $(Y_1, \dots, Y_n)$ be a C-basis of the vector space V of solutions of (E) in L^n.

Definition 6.5.1. *The* differential Galois group *of L over K is the group of differential K-automorphisms of L, that is, the set of automorphisms $\sigma\colon L \to L$ such that*

a) *for any $x \in K$, $\sigma(x) = x$;*
b) *for any $y \in L$, $\sigma(y)' = \sigma(y')$.*

It is denoted $\mathrm{Gal}^D(L/K)$.

Usual Galois groups are subgroups of a group of permutations; similarly, the differential Galois group can be viewed as a subgroup of the group $\mathrm{GL}(V)$ of C-linear automorphisms of the vector space of solutions. (Observe that $\mathrm{GL}(V) \simeq \mathrm{GL}_n(C)$.)

Proposition 6.5.2. *Let $\sigma \in \mathrm{Gal}^D(L/K)$. For any solution Y of the differential equation* (E), *$\sigma(Y)$ is again a solution of* (E) *and the induced map $\sigma|_V\colon V \to V$ so obtained is an isomorphism of C-vector spaces.*

Moreover, the map ρ: $\mathrm{Gal}^D(L/K) \to \mathrm{GL}(V)$ *defined by $\sigma \mapsto \sigma|_V$ is an injective morphism of groups.*

Proof. Let $\sigma \in \mathrm{Gal}^D(L/K)$. Let $Y = (y_1, \dots, y_n)^{\mathrm{t}}$ be a solution of (E). By definition, one then has

$$\sigma(Y)' = \sigma(Y') = \sigma(AY) = A\sigma(Y),$$

since σ is K-linear. Consequently, $\sigma(Y)$ is a solution of (E). This defines a map $\sigma|_V\colon V \to V$, which is obviously C-linear since $C \subset K$. Moreover, $\sigma|_V$ is bijective, for its inverse is given by the restriction of σ^{-1} to V.

It is obvious that the map $\rho\colon \sigma \mapsto \sigma|_V$ is a morphism of groups from $\mathrm{Gal}^D(L/K)$ to $\mathrm{GL}(V)$. It remains to show injectivity. Consider $\sigma \in \mathrm{Gal}^D(L/K)$ with $\rho(\sigma) = \mathrm{id}$; one then has $\sigma(Y_j) = Y_j$ for any j. Since L is generated over K by the coordinates of the Y_j and since $\sigma|_K = \mathrm{id}_K$, one has $\sigma(y) = y$ for any $y \in L$, hence $\sigma = \mathrm{id}$. This shows that ρ is injective. □

Let us now give some examples.

Example 6.5.3 (Exponentials). Let $K = \mathbf{C}(X)$, endowed with its usual derivation and consider the equation $y' = y$. Necessarily, a nonzero solution in a differential extension of $\mathbf{C}(X)$ is transcendental over $\mathbf{C}(X)$. Otherwise, y would be solution of a polynomial equation of minimal degree

$$y^n + a_{n-1}y^{n-1} + \cdots + a_0 = 0, \quad a_0, \ldots, a_{n-1} \in \mathbf{C}(X).$$

Observe that $a_0 \neq 0$. Let us differentiate this relation. One gets

$$ny^{n-1}y' + ((n-1)a_{n-1}y^{n-2}y' + a'_{n-1}y^{n-1}) + \cdots + a'_0 = 0.$$

Hence, since $y' = y$,

$$ny^n + ((n-1)a_{n-1} + a'_{n-1})y^{n-1} + \cdots + (a_1 + a'_1)y + a'_0 = 0,$$

which is another polynomial equation of degree at most n which y satisfies. The two equations must be proportional, hence $a'_0 = na_0$.

However, if $\lambda \in \mathbf{C}^*$, there is no nonzero rational function $R \in \mathbf{C}(X)$ such that $R' = \lambda R$. Let R be a nonzero rational function and let $R = c \prod_{j=1}^{m} P_j{}^{n_j}$ be its factorization as a product of distinct monic irreducible polynomials P_j, with exponents $n_j \in \mathbf{Z} \setminus \{0\}$, and $c \in \mathbf{C}^*$. Then,

$$\frac{R'}{R} = \sum_{j=1}^{m} n_j \frac{P'_j}{P_j}.$$

This is a decomposition into partial fractions, hence R'/R cannot be constant unless $m = 0$, which implies $\lambda = 0$.

This concludes the proof that y is transcendental over $\mathbf{C}(X)$.

The field $\mathbf{C}(X, Y)$ of rational functions in two variables, with the derivation defined by $Y' = Y$, is a Picard-Vessiot extension for this differential equation. By Exercise 6.1, one has, for $P \in \mathbf{C}(X, Y)$, the formula

$$P' = \frac{\partial P}{\partial X} + Y\frac{\partial P}{\partial Y}.$$

The vector space V of solutions has dimension 1, namely $V = \mathbf{C}Y$. An automorphism σ of L which fixes $\mathbf{C}(X)$ is defined by the image of Y, which belongs to V. Therefore, there is some $\rho(\sigma) \in \mathbf{C}^*$ such that $\sigma(Y) = \rho(\sigma)Y$. Conversely, if $c \in \mathbf{C}^*$, the map $\sigma_c \colon P(X, Y) \mapsto P(X, cY)$ defines an automorphism of L, moreover an element of $\operatorname{Gal}^D(L/K)$: if $P \in \mathbf{C}(X, Y)$, one has

$$\begin{aligned} P(X, cY)' &= \frac{\partial P}{\partial X}(X, cY) + (cY)'\frac{\partial P}{\partial Y}(X, cY) \\ &= \frac{\partial P}{\partial X}(X, cY) + cY\frac{\partial P}{\partial Y}(X, cY) \\ &= (\frac{\partial P}{\partial X} + Y\frac{\partial P}{\partial Y})(X, cY) = P'(X, cY), \end{aligned}$$

so that σ_c is a morphism of differential fields.

Finally, the group morphism $\operatorname{Gal}^D(L/K) \to \mathbf{C}^* = GL_1(\mathbf{C})$ is an isomorphism.

Example 6.5.4 (Logarithm). Again let $K = \mathbf{C}(X)$ and consider the differential equation $y' = 1/X$. It is not homogeneous but, as in calculus, its solutions satisfy $(Xy')' = 0$ and we rather study the linear homogeneous equation of second order, $y'' + (1/X)y' = 0$. Let us now find a Picard-Vessiot extension (L, D) for this equation. Any constant is obviously a solution; in particular, $g = 1$ is a solution. Letting $f \in L$ be any nonconstant solution, then (f, g) are independent over $\mathbf{C}$ and necessarily form a basis of the vector space of solutions. Since $Xf'' + f' = (Xf')' = 0$ and $f' \neq 0$, $c = Xf'$ is a nonzero constant, hence we can assume, replacing f by f/c, that $f' = 1/X$.

Let us show that f is transcendental over $\mathbf{C}(X)$. Otherwise, there would be an equation of minimal degree

$$f^n + a_{n-1}f^{n-1} + \cdots + a_0 = 0, \qquad a_0, \ldots, a_{n-1} \in \mathbf{C}(X),$$

which f satisfies. Differentiating, one has

$$nf^{n-1}f' + ((n-1)a_{n-1}f^{n-2}f' + a'_{n-1}f^{n-1}) + \cdots + (a_1 f' + a'_1 f) + a'_0 = 0,$$

which can be rewritten as

$$(n/X + a'_{n-1})f^{n-1} + \cdots + (a_1/X + a'_0) = 0.$$

This is an algebraic relation of degree less than n, hence its coefficients are all zero. In particular, $a'_{n-1} = -n/X$. But, if $\lambda \in \mathbf{C}^*$, there is no rational function $R \in \mathbf{C}(X)$ such that $R' = \lambda/X$. Otherwise, we may write the decomposition into partial fractions of R as

$$R = P + \sum_{j=1}^{m} \frac{Q_j}{P_j^{n_j}},$$

for some polynomial P, some integer m, and where for any j with $1 \leqslant j \leqslant m$, P_j is a monic irreducible polynomial, Q_j a polynomial which is prime to P_j and whose degree is less than $n_j \deg(P_j)$. Then

$$R' = P' + \sum_{j=1}^{m} \frac{Q'_j P_j - n_j P'_j Q_j}{P_j^{n_j+1}},$$

which is the decomposition of R' into partial fractions, except that some terms may be zero. By uniqueness of such a decomposition, one finds that

– $P' = 0$, hence P is constant;

– for any j with $P_j \neq X$, $Q'_j P_j - n_j P'_j Q_j = 0$. Since P_j is irreducible and since it does not divide P'_j, P_j must divide Q_j, which is a contradiction;

– finally, if $P_j = X$, then $Q'_j X - n_j Q_j = -nX^m$. Since $n_j \geqslant 1$, X divides Q, again a contradiction.

Consequently, R is constant and $R' = 0 \neq \lambda/X$ since $\lambda \neq 0$. This concludes the proof that f is transcendental over $\mathbf{C}(X)$.

It follows that $L = \mathbf{C}(X,Y)$, endowed with the derivation defined by $Y' = 1/X$, is a Picard-Vessiot extension for the equation $xy'' + y' = 0$. By Exercise 6.1, the derivative of $P \in L$ is given by the formula

$$P' = \frac{\partial P}{\partial X} + \frac{1}{X}\frac{\partial P}{\partial Y}.$$

The basis (f,g) of the space of solutions allows us to identify V with $\mathbf{C}^2$ and $\mathrm{GL}(V)$ with $\mathrm{GL}_2(\mathbf{C})$ Let $\sigma \in \mathrm{Gal}^D(L/K)$ be any differential automorphism of L. Since $g = 1$ belongs to K, one has $\sigma(g) = g$. Moreover, there are a and $b \in \mathbf{C}$ such that $\sigma(f) = af + bg$. Deriving, one finds $\sigma(f)' = af' = a/X$, but $\sigma(f)' = \sigma(f') = \sigma(1/X) = 1/X$. Consequently, $a = 1$, and the image of the homomorphism $\mathrm{Gal}^D(L/K) \to \mathrm{GL}_2(\mathbf{C})$ is contained in the subgroup U of 2×2 matrices of the form $\left(\begin{smallmatrix} 1 & 0 \\ b & 1 \end{smallmatrix}\right)$, for $b \in \mathbf{C}$.

Conversely, if $c \in \mathbf{C}$, the map $\sigma\colon P \mapsto P(X, Y+c)$ defines a differential automorphism, for, if $P \in \mathbf{C}(X,Y)$,

$$\begin{aligned}
\sigma(P)' &= P(X,Y+c)' \\
&= \frac{\partial P}{\partial X}(X,Y+c) + (Y+c)'\frac{\partial P}{\partial Y}(X,Y+c) \\
&= \Big(\frac{\partial P}{\partial X} + \frac{1}{X}\frac{\partial P}{\partial Y}\Big)(X,Y+c) \\
&= \sigma(P').
\end{aligned}$$

Therefore, $\mathrm{Gal}^D(L/K) \simeq U$. It must also be observed that the map $c \mapsto \left(\begin{smallmatrix} 1 & 0 \\ c & 1 \end{smallmatrix}\right)$ is an isomorphism of groups, $\mathbf{C} \simeq U$.

Example 6.5.5 (Galois extensions). Let (K,D) be any differential field. Assume that its constant field C is algebraically closed and of characteristic 0, Let $K \to L$ be any Galois extension of K. By Theorem 6.2.6, there is a unique derivation D on L such that the morphism $K \to L$ is a morphism of differential fields. Let us show that this is a Picard-Vessiot extension for some differential equation.

By Lemma 6.4.4, a constant in L, being algebraic over K, is algebraic over the field of constants of K, which is C. Since C is algebraically closed, C is also the field of constants in L.

On the other hand, let $n = [L : K]$ and let f be any element of L. The dimension of the K-vector space generated by $f, f', f'', \dots$ in L is finite, less or equal than n. This implies that f satisfies a nontrivial differential equation.

Let σ be an element of $\mathrm{Gal}(L/K)$, and let us consider the map $\tilde{D}\colon L \to L$ defined by $\tilde{D}(x) = \sigma(D(\sigma^{-1}(x)))$. This is a derivation of L. First of all, this is a morphism of abelian groups. Moreover, if x, $y \in L$, one has

$$\begin{aligned}\tilde{D}(xy) &= \sigma(D(\sigma^{-1}(xy))) = \sigma(D(\sigma^{-1}(x)\sigma^{-1}(y)))\\ &= \sigma\big(D(\sigma^{-1}(x))\,\sigma^{-1}(y) + D(\sigma^{-1}(x))\,\sigma^{-1}(y)\big)\\ &= \sigma(D(\sigma^{-1}(x)))\sigma(\sigma^{-1}(y)) + \sigma(D(\sigma^{-1}(y)))\sigma(\sigma^{-1}(x))\\ &= \tilde{D}(x)y + \tilde{D}(y)x.\end{aligned}$$

Since D is the only derivation of L extending the derivation of K, one has $\tilde{D} = D$, which means that for any $x \in L$, $\sigma(D(x)) = D(\sigma(x))$. In other words, elements of $\mathrm{Gal}(L/K)$ are elements of $\mathrm{Gal}^D(L/K)$ and the natural injection $\mathrm{Gal}^D(L/K) \to \mathrm{Gal}(L/K)$ is an isomorphism. Moreover, any differential equation satisfied by f is also satisfied by $\sigma(f)$.

Since the extension $K \to L$ is Galois, we may consider some $f \in L$ which is a primitive element. Let V denote the C-vector space in L generated by the conjugates of f and let $(f_1, \dots, f_d)$ be a basis of V consisting of conjugates of f.

Let us construct a differential equation of order d whose space of solutions is V. By construction, $f_1, \dots, f_d$ are linearly independent over C, so their Wronskian $W(f_1, \dots, f_d)$ does not vanish. On the other hand, introducing a formal differential variable and expanding the Wronskian determinant

$$W(f_1, \dots, f_d, Y) = \begin{vmatrix} f_1 & \dots & f_d & Y \\ f_1' & \dots & f_d' & Y' \\ \vdots & & \vdots & \vdots \\ f_1^{(d)} & \dots & f_d^{(d)} & Y^{(d)} \end{vmatrix}$$

along the first column, one finds elements $A_0, \dots, A_d \in L$ such that

$$W(f_1, \dots, f_d, Y) = A_0 Y + A_1 Y' + \dots + A_d Y^{(d)}.$$

Note that $A_d = W(f_1, \dots, f_d) \neq 0$.

We shall in fact show that for any j, $A_j/A_d \in K$, so that the differential equation

$$Y^{(d)} + \frac{A_{d-1}}{A_d} Y^{(d-1)} + \dots + \frac{A_0}{A_d} Y = 0 \tag{E}$$

has its coefficients in K. We have to show that for any j with $0 \leqslant j \leqslant d$, and any $\sigma \in \mathrm{Gal}(L/K)$, $\sigma(A_j)/\sigma(A_d) = A_j/A_d$.

To do that, notice that $W(f_1, \dots, f_d, Y)$ depends only on the vector space V. More precisely if $P(\mathbf{f}, \mathbf{g}) \in \mathrm{GL}_d(C)$ is the matrix sending a basis $\mathbf{g} = (g_1, \dots, g_d)$ of V to the basis $\mathbf{f} = (f_1, \dots, f_d)$, one has

$$W(g_1, \dots, g_d, Y) = \det P(\mathbf{f}, \mathbf{g}) W(f_1, \dots, f_d, Y).$$

If σ is any element in $\mathrm{Gal}(L/K)$, $\sigma(f_1), \dots, \sigma(f_d)$ form a basis $\sigma(\mathbf{f})$ of V, so that

$$W(\sigma(f_1),\dots,\sigma(f_d),Y) = \det P(\mathbf{f},\sigma(\mathbf{f}))\, W(f_1,\dots,f_d,Y).$$

Expanding the determinant, we find that for any $j \in \{0,\dots,d\}$, $\sigma(A_j) = \det P(\mathbf{f},\sigma(\mathbf{f}))A_j$. Consequently, $\sigma(A_j/A_d) = A_j/A_d$, as was to be shown.

The space of solutions of (E) in L is equal to V, hence has dimension d. Moreover, L is generated by f_1 as a field, thus it is generated by V as a differential field. Finally, L has no constant elements other than the elements of C. It follows that it is a Picard-Vessiot extension of (E).

Moreover, viewed in matrix form, the isomorphism $\mathrm{Gal}^D(L/K) \to \mathrm{Gal}(L/K)$ corresponds essentially to the (inverse of the) classical morphism from the symmetric group $\mathfrak{S}_n$ to the linear group $\mathrm{GL}_n(C)$ given by permutation matrices.

6.6 The differential Galois correspondence

In the algebraic theory of differential equations, there are Galois groups, and also an analogue of Galois correspondence. Proofs are too difficult to be given here, but I would like to convey some ideas about the *statements.*

Fix a Picard-Vessiot extension $K \subset L$, with field of constants C, assumed algebraically closed and of characteristic zero, corresponding to a differential equation $Y' = AY$, for some $A \in M_n(K)$. We saw that its differential Galois group $\mathrm{Gal}^D(L/K)$ can be seen naturally as a subgroup of $\mathrm{GL}_n(C)$.

For any subgroup $H \subset \mathrm{Gal}^D(L/K)$, we can introduce the subfield L^H consisting of all $x \in L$ such that $\sigma(x) = x$ for any $\sigma \in H$. It is easy to check that the derivation $D\colon L \to L$ maps L^H into itself, so that L^H is a differential field, and the extension $K \subset L^H$ is a differential extension, as is the extension $L^H \subset L$.

There is, however, a new feature: differential Galois groups are not mere subgroups of some $\mathrm{GL}_n(C)$. They are automatically *algebraic groups*, which means that there are polynomials $P_j \in C[a_{11},\dots,a_{nn}]$ such that a matrix $A = (a_{ij}) \in \mathrm{GL}_n(C)$ belongs the differential Galois group if and only if $P_j(a_{11},\dots,a_{nn}) = 0$ for all j.

Hence, the basic results in differential Galois theory are as follows.

– For any algebraic subgroup $H \subset \mathrm{Gal}^D(L/K)$, the extension $L^H \subset L$ is a Picard-Vessiot extension (for some differential equation), and its differential Galois group can be identified with H.

– Conversely, any differential subextension $K \subset E$ is of the form L^H for some algebraic subgroup $H \subset \mathrm{Gal}^D(L/K)$.

– If $H \subset \mathrm{Gal}^D(L/K)$ is an algebraic subgroup, then the differential extension $K \subset L^H$ is a Picard-Vessiot extension if and only if H is a normal subgroup of $\mathrm{Gal}^D(L/K)$. Then, $\mathrm{Gal}^D(L^H/K) \simeq \mathrm{Gal}^D(L/K)/H$.

Concerning the analogue of solvability by radicals, one is interested in solving a differential equation by *quadratures*, that is, using only indefinite integrals. So that the theory is neat, one also needs to authorize algebraic extensions. Then, one can prove that a differential equation is solvable by quadratures and algebraic extensions if and only if the connected component of the identity[1] of $\mathrm{Gal}^D(L/K)$ is solvable. This is where Lie-Kolchin's theorem 4.7.2 intervenes in the algebraic theory of differential equations.

6.7 Integration in finite terms, elementary extensions

In this last section, I want to prove a theorem of Liouville's concerning functions of which an antiderivative can be computed "in finite terms," for example, using only logarithms or exponentials. Liouville had proved such results through a series of articles published around 1830, but it was not until the middle of 20th century that Ostrowski recasted Liouville's theorem in the algebraic setting of differential fields. The theory has evolved to an *algorithm* (Risch) to compute indefinite integrals, at least when possible; it is implemented in most computer algebra systems.

Joseph Liouville (1809–1882)

All fields are assumed to have characteristic zero.

Definition 6.7.1. *Let (K, D) be a differential field and let $a \in K$. One says an element t in a differential extension of K is a* logarithm *of the element a if $a \neq 0$ and if $t' = a'/a$. One says that t is an* exponential *of a if $t \neq 0$ and $t'/t = a'$.*

Definition 6.7.2. *Let (K, D) be a differential field. A differential extension (L, D) of (K, D) is said to be* elementary *if there are elements $t_1, \ldots, t_n$ in L such that*

a) *$L = K(t_1, \ldots, t_n)$;*
b) *the field of constants of L is equal to that of K;*

and such that for any j, one of the three following properties holds:

[1] If $C = \mathbf{C}$ is the field of complex numbers, this is the connected component in the usual, topological, sense. For an arbitrary field C, one needs to consider the so-called Zariski topology.

3a) *t_j is algebraic over $K(t_1, \dots, t_{j-1})$;*
3b) *t_j is a logarithm of a nonzero element in $K(t_1, \dots, t_{j-1})$;*
3c) *t_j is an exponential of an element in $K(t_1, \dots, t_{j-1})$.*

Exercise 6.7.3. Check that, L being an elementary differential extension of K as in the previous definition, the subfields $K(t_1, \dots, t_j)$, with $1 \leqslant j \leqslant n$, are differential subfields of L.

Theorem 6.7.4 (Liouville, 1835; Ostrowski, 1946). *Let (K, D) be a differential field of characteristic zero and let $f \in K$. If f has an antiderivative in an elementary differential extension of K, then there exists an integer $n \geqslant 0$, constants $c_1, \dots, c_n \in K$ and elements $u_1, \dots, u_n, v$ in K such that*

$$f = v' + \sum_{i=1}^{n} c_i \frac{u_i'}{u_i}.$$

The proof is done by induction on the number of steps in Definition 6.7.2 of an elementary extension. One therefore needs a lemma concerning "one-step" elementary differential extensions, that is, differential extensions generated by either an algebraic element, an exponential or a logarithm, with the same field of constants. Assuming that we have proved the proposition below, Theorem 6.7.4 is proven as follows. We begin with $n = 0$, by taking for v an antiderivative of f. Using the proposition below, we successively follow each step in the definition of an elementary extension. The final step gives us precisely Liouville's theorem.

Proposition. *Let $K \subset K(t)$ a "one-step" elementary differential extension and let $f \in K$ which can be written as*

$$f = v' + \sum_{i=1}^{n} c_i \frac{u_i'}{u_i}$$

for some constants c_i and elements $u_1, \dots, u_n$ and v in $K(t)$. Then, f admits a similar expression in K.

We prove this proposition by distinguishing three cases.

First case: t is algebraic over K. Let $K \subset L$ be a Galois closure of the algebraic extension $K \subset K(t)$, and endow L with its unique derivation such that $K \to L$ is a morphism of differential fields. If $\sigma \in \mathrm{Gal}(L/K)$ and $x \in L$, $x \neq 0$, one has

$$\sigma\left(\frac{x'}{x}\right) = \frac{\sigma(x')}{\sigma(x)} = \frac{\sigma(x)'}{\sigma(x)}.$$

Then compute

$$[L:K]f = \sum_{\sigma\in\operatorname{Gal}(L/K)} \sigma(f) = \sum_{\sigma\in\operatorname{Gal}(L/K)} \sigma(v)' + \sum_{i=1}^{n} c_i \sum_{\sigma\in\operatorname{Gal}(L/K)} \sigma(u_i'/u_i).$$

Set $\tilde{v} = (\sum_\sigma \sigma(v))/[L:K]$ and $\tilde{u}_i = \prod_\sigma \sigma(u_i)$. These elements of L are invariant under the action of $\operatorname{Gal}(L/K)$, hence belong to K. Moreover,

$$f = \tilde{v}' + \sum_{i=1}^{n} \frac{1}{[L:K]} c_i \frac{\tilde{u}_i'}{\tilde{u}_i}.$$

Second case: t is transcendental over K, and is a logarithm. Then we can identify the field $K(t)$ with the field of rational functions in t, although with another structure of a differential field. Let π be any monic irreducible polynomial in $K[T]$. For any $U \in K(T)^*$, define $\operatorname{ord}_\pi(U)$ to be the exponent of π when one writes U as a product of distinct monic irreducible polynomials of $K[T]$, times an element in K^*. If $u = U(t) \in K(t)$, we write $\operatorname{ord}_\pi(u)$ for $\operatorname{ord}_\pi(U)$. Similarly, if $u = U(t)$ for $U \in K[T]$, we write $\deg u$ for $\deg U$.

Lemma 6.7.5. (Assuming $t' \in K^*$.) *Let $u \in K(t)^*$ and let π be any irreducible monic polynomial in $K[T]$.*

a) *If* $\operatorname{ord}_\pi(u) = 0$, *then* $\operatorname{ord}_\pi(u'/u) \geqslant 0$.
b) *If* $\operatorname{ord}_\pi(u) \neq 0$, *one has* $\operatorname{ord}_\pi(u'/u) = -1$.
c) *If $u = U(t)$ for some $U \in K[T]$, then* $\deg u - 1 \leqslant \deg u' \leqslant \deg u$.

Proof of the lemma. Let $U \in K(T)$ with $u = U(T)$ and write $U = a \prod_j \pi_j(T)^{n_j}$, for $a \in K^*$, nonzero integers $n_j \in \mathbf{Z}$, and distinct irreducible monic polynomials $\pi_j \in K[T]$. One thus has $u = a \prod_j \pi_j(t)^{n_j}$ and

$$\frac{u'}{u} = \frac{a'}{a} + \sum_j n_j \frac{\pi_j(t)'}{\pi_j(t)} = \frac{a'}{a} + \sum_j n_j \frac{\pi_j^D(t) + t'\pi_j'(t)}{\pi_j(t)}.$$

Now, for any j, $\pi_j^D + t'\pi_j'$ is a polynomial of $K[T]$ of degree $< \deg(\pi_j)$, since π_j is monic. Moreover, t is transcendental over K, so that $\pi_j(t) \notin K$, hence $\pi_j(t)' \neq 0$ because $K(t)$ and K are assumed to share the same field of constants. Consequently, π_j does not divide the polynomial $\pi_j^D + t'\pi_j'$ and $\operatorname{ord}_{\pi_j}$ of the jth term in u'/u is equal to -1. Since $\operatorname{ord}_{\pi_j}$ of the other terms is 0, one has $\operatorname{ord}_{\pi_j}(u'/u) = -1$, as claimed by *b*). The formula above also shows that for any irreducible monic polynomial π, not among the π_j, one has $\operatorname{ord}_\pi(u'/u) \geqslant 0$, hence *a*).

Let us now show *c*). Let $u = U(t)$ for $U = u_0 + u_1 T + \cdots + u_n T^n \in K[T]$, with $u_n \neq 0$ so that $\deg u = n$. One has

$$u' = (u_0' + u_1 t') + (u_1' + 2u_2 t')t + \cdots + (u_{n-1}' + nu_n t')t^{n-1} + u_n' t^n.$$

If $u_n' \neq 0$, then $\deg(u') = n = \deg(u)$. Otherwise, observe that the vanishing of

$$u_{n-1}' + nu_n t' = (u_{n-1} + nu_n t)'$$

implies that $u_{n-1} + nu_n t$ is constant, hence is in K, hence $t \in K$, which is absurd! Consequently, $u_{n-1}' + nu_n t' \neq 0$ and $\deg(u') = \deg(u) - 1$. □

Going back to the proof of the second case in the proof of the proposition, let us expand the logarithmic derivatives in the formula

$$f = v' + \sum_i c_i \frac{u_i'}{u_i},$$

so that it can be rewritten as

$$f = v' + \sum_i c_i \frac{a_i'}{a_i} + \sum_\pi c_\pi \frac{\pi'}{\pi}, \qquad (*)$$

where the a_i belong to K^*, the π are monic irreducible polynomials in $K[T]$, and c_i, c_π are constants in K.

For any irreducible monic polynomial $\pi \in K[T]$ appearing in the denominator of v, one has $\operatorname{ord}_\pi(v') = \operatorname{ord}_\pi(v) - 1 \leqslant -2$. However, ord_π of any other terms is $\geqslant -1$, which makes it impossible for the whole sum to be equal to f, for we have $\operatorname{ord}_\pi(f) = 0$. This shows that $v = V(t)$ for $V \in K[T]$.

Since $\deg \pi(t)' < \deg \pi(t)$, equation $(*)$ is a decomposition of the "constant" rational function f into partial fractions. Since such a decomposition is unique, polar terms vanish, and one has

$$f = v' + \sum_i c_i \frac{a_i'}{a_i}.$$

In particular $v' = f - \sum_i c_i(a_i'/a_i)$ belongs to K, so that its degree is zero. This implies that $V(T) = cT + d$ for c and $d \in K$, and $c' = 0$. It follows that

$$f = ct' + d' + \sum_i c_i \frac{a_i'}{a_i} = d' + \left(c\frac{a'}{a} + \sum_i c_i \frac{a_i'}{a_i} \right)$$

if t is the logarithm of $a \in K^*$, whence $t' = a'/a$. This is an expression of the expected form, hence the proposition in this case.

Third case: t is transcendental over K, and is an exponential.

We continue to identify elements of $K(t)$ with rational functions in one variable T. The proof will be similar to that of the previous case, using the following lemma instead of Lemma 6.7.5.

Lemma 6.7.6. (Assuming $t'/t \in K^*$.) *Let $\pi \in K[T]$ be any irreducible monic polynomial and let $u \in K(t)$.*

a) *If* $\operatorname{ord}_\pi(u) = 0$, *then* $\operatorname{ord}_\pi(u'/u) \geqslant 0$.

b) *Assume* $\operatorname{ord}_\pi(u) \neq 0$. *If $\pi \neq T$, then* $\operatorname{ord}_\pi(u'/u) = -1$; *if $\pi = T$, then* $\operatorname{ord}_\pi(u'/u) \geqslant 0$.

c) *If $u = U(t)$ with $U \in K[T]$, then* $\deg(u') = \deg(u)$.

Proof of the lemma. We begin as in the proof of Lemma 6.7.5: if $u = a\prod_j \pi_j(t)^{n_j}$, one has

$$\frac{u'}{u} = \frac{a'}{a} + \sum_j n_j \frac{\pi_j(t)'}{\pi_j(t)} = \frac{a'}{a} + \sum_j n_j \frac{\pi_j^D(t) + t'\pi_j'(t)}{\pi_j(t)}.$$

Assertion a) follows from this formula. Moreover, the degree of the polynomial $\pi_j^D(T) + (t'/t)T\pi_j'(T)$ is less or equal than the degree of π_j. This implies that either these two polynomials are coprime, and $\operatorname{ord}_{\pi_j}(u'/u) = -1$, or there exists $\lambda \in K$ such that $\pi_j^D + (t'/t)T\pi_j' = \lambda\pi_j$, and $\operatorname{ord}_{\pi_j}(u'/u) \geqslant 0$.

Let us now show that this second case happens only for $\pi_j = T$. Write $\pi = T^n + p_{n-1}T^{n-1} + \cdots + p_0$, with $p_0, \ldots, p_{n-1} \in K$. Then, denoting $a = t'/t$, the relation $\pi^D + aT\pi' = \lambda\pi$ can be rewritten as

$$\begin{aligned} anT^n + (p_{n-1}' + a(n-1)p_{n-1})T^{n-1} + \cdots + (p_1' + ap_1)T + p_0' \\ = \lambda T^n + \lambda p_{n-1}T^{n-1} + \cdots + \lambda p_0. \end{aligned}$$

Consequently, $\lambda = an$. Moreover, for any integer j such that $0 \leqslant j < n$ and $p_j \neq 0$, one has $p_j'/p_j = a(n-j) = (n-j)t'/t$. This implies that t^{n-j}/p_j is constant, hence is in K, which contradicts the fact that t is transcendental over K. It follows that $\pi = T^n$, hence $\pi = T$ since π is irreducible. Conversely, if $\pi = T$, one has $\pi(t)'/\pi(t) = t'/t \in K^*$, hence $\operatorname{ord}_\pi(u'/u) \geqslant 0$.

To prove c), let us consider $u = U(T)$ for some polynomial $U \in K[T]$. If $a = t'/t \in K^*$ and $U = u_nT^n + \cdots + u_0$, with $u_n \neq 0$, one has

$$u' = \sum_{k=0}^{n} (u_k' + aku_k)t^k,$$

so that $\deg(u') \leqslant \deg(u)$. If one had $u_n' + nau_n = 0$, then

$$\frac{u_n'}{u_n} = -na = -n\frac{t'}{t},$$

so that u_nt^n would be a constant, hence in K. However, this contradicts the assumption that t is transcendental over K. Consequently, $u_n' + nau_n \neq 0$ and $\deg(u') = \deg(u)$. □

We now go back to the proof of the proposition in the case where t is an exponential, that is, $t' = a't$, for some $a \in K^*$. As above, we expand the logarithmic derivatives in an expression

$$f = v' + \sum c_i \frac{u_i'}{u_i},$$

to get

$$f = v' + \sum_i c_i \frac{a_i'}{a_i} + \sum_\pi c_\pi \frac{\pi'}{\pi}, \qquad (**)$$

where the a_i belong to K^*, π runs over monic irreducible polynomials in $K[T]$, and c_i, c_π are nonzero constants in K. Consequently, one has

$$f = v' + \sum_i c_i \frac{a_i'}{a_i} + \sum_\pi c_\pi a' + \sum_\pi c_\pi \frac{\pi^D(t) + a't\pi'(t) - a'\pi(t)}{\pi(t)}.$$

If $\pi \neq T$ and $\operatorname{ord}_\pi(v) < 0$, then $\operatorname{ord}_\pi(v') \leqslant -2$, although ord_π of the right-hand side is at least -1. This shows that $\operatorname{ord}_\pi(v) \geqslant 0$ except maybe for $\pi = T$; in other words, one can write $v = \sum_{j\in\mathbf{Z}} v_j t^j$, for some $v_j \in K$. Then, observe that for any π, $\deg(\pi^D + a'T\pi' - a'\pi) < \deg(\pi)$. Since $f \in K$ is a "constant" rational function, uniqueness of the decomposition into partial fractions implies that we can omit the last sums (those with π in the denominator). Denoting $c = \sum_\pi c_\pi$, equation $(**)$ now becomes

$$f = \sum_j (v_j' + a'jv_j)t^j + ca' + \sum_i c_i \frac{a_i'}{a_i}.$$

Comparing the degree 0 terms, one gets

$$f = (v_0 + ac)' + \sum_i c_i \frac{a_i'}{a_i},$$

which is an expression of the form required by the proposition.

Having concluded the proof of the last case of the intermediate proposition, the proof of Liouville's theorem 6.7.4 is completed. Let us now give some concrete applications.

Proposition 6.7.7. *Let f and g be two rational functions in $\mathbf{C}(X)$. Assume that $f \neq 0$, that g is not constant, and that $f\exp(g)$ has an antiderivative in an elementary differential extension of $\mathbf{C}(X, \exp(g))$. Then, there exists a in $\mathbf{C}(X)$ such that*

$$f = a' + ag'.$$

Proof. First, we observe that if $f = a' + ag'$, then one has $f\exp(g) = (a\exp(g))'$. Conversely, assume that f has an elementary antiderivative. By Liouville's theorem, we may write

$$f\exp(g) = v' + \sum_{i=1}^{n} c_i \frac{u_i'}{u_i},$$

for $v \in \mathbf{C}(X, \exp(g))$, $c_i \in \mathbf{C}$, $u_i \in \mathbf{C}(X, \exp(g))$. Set $T = \exp(g)$; it is transcendental over $\mathbf{C}(X)$ (see Exercise 1.3), so that we can identify u_i and v to rational functions in one variable T with coefficients in the field $\mathbf{C}(X)$. As above, we may assume that either $u_i \in \mathbf{C}(X)$, or u_i is an irreducible monic polynomial with coefficients in $\mathbf{C}(X)$.

As in the proof of the last case of Liouville's theorem, the only u_i which can appear are T or elements of $\mathbf{C}(X)$. Similarly, the denominator of v is a power of T and we write $v = \sum_j v_j(X)T^j$ (the j are rational integers), hence

$$fT = \sum_j (v_j' + jg'v_j)T^j + cg' + \sum_i c_i \frac{u_i'(X)}{u_i(X)}.$$

Comparing the coefficients of T on each side, we find

$$f = v_1' + g'v_1,$$

hence the proposition, with $a = v_1$. □

Example 6.7.8. The "function" $\exp(x^2)$ *has no elementary antiderivative.* Otherwise, there would exist $a \in \mathbf{C}(x)$ such that $1 = a' + 2xa$. But this is impossible: a pole of a is a double pole of a', and an at most simple pole of $1 - 2xa$, so that a is a polynomial. Then $1 - a' = 2xa$ although they do not have the same degree.

Exercise 6.6 proposes other explicit examples.

6.8 Appendix: Hilbert's Nullstellensatz

This section is devoted to the study of the maximal ideals in the ring $k[X_1, \ldots, X_n]$ of polynomials in n variables with coefficients in a field k. The main result is Hilbert's Nullstellensatz (literally, theorem of the location of zeros). This result has many incarnations, all interesting. Here are three of them. When k is uncountable, for example, in the already very important case $k = \mathbf{C}$, one can give a very simple proof of these theorems.

David Hilbert (1862–1943)

Theorem 6.8.1 (Algebras). *Let k be a field and let A be any finitely generated k-algebra If A is a field, then it is a finite (algebraic) extension of k.*

Theorem 6.8.2 (Ideals). *Let k be an algebraically closed field. Let I be a maximal ideal in the ring $k[X_1, \ldots, X_n]$. Then there are $a_1, \ldots, a_n$ in k such that $I = (X_1 - a_1, \ldots, X_n - a_n)$.*

Theorem 6.8.3 (Equations). *Let k be an algebraically closed field and let $P_1, \ldots, P_m$ be polynomials of $k[X_1, \ldots, X_n]$. If the system of algebraic equations*

$$P_1(x_1, \ldots, x_n) = \cdots = P_m(x_1, \ldots, x_n) = 0$$

has no solution in k^n, then there are polynomials $Q_1, \ldots, Q_m$ such that

$$1 = P_1 Q_1 + \cdots + P_m Q_m.$$

This last form of Hilbert's Nullstellensatz means that if a system of polynomial equations in many variables has no solution in an algebraically closed field, the system is incompatible in a very strong sense, for an equality as given by the theorem obviously prevents the system $P_1 = \cdots = P_m = 0$ from having solutions in *any* field containing k.

Proof of Theorem 6.8.1 when k is uncountable. Let $x_1, \ldots, x_n$ in A such that $A = k[x_1, \ldots, x_n]$. Any element of A can be written (not uniquely) as a polynomial in $x_1, \ldots, x_n$. This implies that the dimension of A as a k-vector space is not greater than the cardinality of the set of monomials in n variables. In particular, $\dim_k A$ is at most countable.

Assume now that one of the x_j, say x_1, is transcendental over k. Then, the ring $k[x_1]$ is isomorphic to the ring of polynomials $k[X]$, and A being a field, A contains the field $k(x_1)$ which is isomorphic to the field of rational functions $k(X)$. However, the elements $1/(X-a)$, for $a \in k$, are linearly independent over k. This can be proved directly, but also follows from the uniqueness of the decomposition of rational functions into partial fractions. Since k is uncountable, the dimension of $k(X)$ as a k-vector space is uncountable. This contradicts the existence of an inclusion $k(X) \subset A$.

Consequently, $x_1, \dots, x_n$ are algebraic over k and A is algebraic over k. By induction, one even sees that the extension $k \subset A$ is finite. □

Proof of Theorem 6.8.2. Let $(a_1, \dots, a_n) \in k^n$ and let I be the ideal of $k[X_1, \dots, X_n]$ generated by $X_1 - a_1, \dots, X_n - a_n$. Let $P \in k[X_1, \dots, X_n]$ and write down the Euclidean division of P by $X_1 - a_1$ (the degree variable being X_1). Thus there exist polynomials P_1 and $Q_1 \in k[X_1, \dots, X_n]$ such that

$$P(X_1, \dots, X_n) = (X_1 - a_1)Q_1(X_1, \dots, X_n) + P_1(X_1, \dots, X_n),$$

the polynomial P_1 being of degree $< \deg(X_1 - a_1)$ in the variable X_1. This means that P_1 does not depend on X_1, hence $P_1 \in k[X_2, \dots, X_n]$. By induction, we find an expression

$$P(X_1, \dots, X_n) = (X_1 - a_1)Q_1 + (X_2 - a_2)Q_2 + \dots + (X_n - a_n)Q_n + P_n,$$

where P_n is a constant polynomial, necessarily equal to $P(a_1, \dots, a_n)$. It follows that $P(a_1, \dots, a_n) = 0$ if and only if $P \in I$.

Let J be any ideal of $k[X_1, \dots, X_n]$ containing I, with $J \neq I$, and let $P \in J \setminus I$. One has $P(a_1, \dots, a_n) \neq 0$. Since the polynomial

$$P(X_1, \dots, X_n) - P(a_1, \dots, a_n)$$

vanishes at $(a_1, \dots, a_n)$, it belongs to I, so to J. It follows that $P(a_1, \dots, a_n)$ belongs to J. Since this is a nonzero element of k, its inverse belongs to J too and $1 \in J$, so that $J = k[X_1, \dots, X_n]$. This shows that I is a maximal ideal.

Conversely, let I be any maximal ideal of $k[X_1, \dots, X_n]$. Let us introduce the quotient ring $A = k[X_1, \dots, X_n]/I$ and the canonical morphism $k[X_1, \dots, X_n] \to A$. Denote by x_j the image of X_j in A. One has $A = k[x_1, \dots, x_n]$ and A is a finitely generated k-algebra, and so is A/I. Since I is a maximal ideal, A/I is a field. By Theorem 6.8.1, the extension $k \subset A$ is algebraic. Since k is algebraically closed, this is an isomorphism and there is for any j an element $a_j \in k$ such that $x_j = a_j$. In other words, $X_j - a_j \in I$ for any j and the ideal I contains the maximal ideal $(X_1 - a_1, \dots, X_n - a_n)$. One must have equality, q.e.d. □

Proof of Theorem 6.8.3. Let I be the ideal generated by $P_1, \dots, P_m$. Let $\mathfrak{m}$ be any maximal ideal of $k[X_1, \dots, X_n]$. If one had $I \subset \mathfrak{m}$, the n-tuple

$(a_1, \ldots, a_n)$ such that $\mathfrak{m} = (X_1 - a_1, \ldots, X_n - a_n)$ would be a solution of the system of equations $P_j(x_1, \ldots, x_n) = 0$, $1 \leqslant j \leqslant m$. This means that no maximal ideal of $k[X_1, \ldots, X_n]$ contains I, so that $I = k[X_1, \ldots, X_n]$ by Krull's theorem 2.5.3. Therefore, $1 \in I$ and there are polynomials $Q_1, \ldots, Q_m$ such that $1 = P_1 Q_1 + \cdots + P_m Q_m$. □

Exercises

Exercise 6.1. **a)** Let (A, D) be a differential ring, and let $P \in A[X, Y]$ be a polynomial in two variables X and Y. Denote by P^D the polynomial obtained by differentiating the coefficients of P, and by $\partial P/\partial X$ et $\partial P/\partial Y$ the partial derivatives of P with respect to X and Y.

For any elements x, $y \in A$, show that

$$P(x, y)' = P^D(x, y) + x' \frac{\partial P}{\partial X}(x, y) + y' \frac{\partial P}{\partial X}(x, y).$$

b) Extend this formula to the case of rational fractions $P \in K(X, Y)$, where K is a differential field.

Exercise 6.2. Let (A, D) be any differential ring and let $G \in M_n(A)$. Show that

$$(\deg G)' = \operatorname{Tr}(\operatorname{Com}(G) G'),$$

where $\operatorname{Com}(G)$ denotes the comatrix of G, that is, the transpose of the matrix of cofactors.

Exercise 6.3. Let A be any ring. Denote by $A[\varepsilon]$ the ring $A[X]/(X^2)$, where ε denotes the class of X. In other words, $A[\varepsilon] = \{a + b\varepsilon\,;\, a, b \in A\}$, addition and multiplication being given by $(a + b\varepsilon) + (a' + b'\varepsilon) = (a + a') + (b + b')\varepsilon$ and $(a + b\varepsilon)(a' + b'\varepsilon) = (aa') + (a'b + ab')\varepsilon$.

Let π denote the morphism of rings, $\pi\colon A[\varepsilon] \to A$, given by $\pi(a + b\varepsilon) = a$ for any a and $b \in A$

a) Let D be any derivation of A. Show that the map $\varphi_D\colon A \to A[\varepsilon]$ defined by $\varphi_D(a) = a + \varepsilon D(a)$ is a ring morphism.

b) Conversely, if $\varphi\colon A \to A[\varepsilon]$ is a ring morphism with $\pi \circ \varphi = \mathrm{id}_A$, show that there is a unique derivation D of A such that $\varphi = \varphi_D$.

Exercise 6.4. The goal of this exercise is to describe a Picard-Vessiot extension for the equation $y'' + y = 0$, over the field $\mathbf{C}(X)$ of rational functions, endowed with the usual derivation.

Define $R = \mathbf{C}[X, Y_1, Y_2, Y_1', Y_2']$ and endow it with the unique derivation such that $(Y_j)' = Y_j'$ and $(Y_j')' = -Y_j$.

a) Show that the set of constant elements of R contains $\mathbf{C}$, but contains other elements. (Think of the trigonometric relation $\sin^2(x) + \cos^2(x) = 1$.)

b) Show that the field $\mathbf{C}(X, Y)$ endowed with the derivation defined by $Y' = iY$ is a Picard-Vessiot extension of the equation $y'' + y = 0$.

Exercise 6.5. Let (K, D) be a differential field with field of constants C, which we assume to be algebraically closed and of characteristic zero.

Let (E): $y^{(n)} + a_{n-1}y^{(n-1)} + \cdots + a_0 y = 0$ be any order n differential equation. Let (L, D) be a Picard-Vessiot extension for this equation.

Let $f_1, \ldots, f_n$ be a C-basis of the vector space of solutions of (E) in L.

a) Compute the derivative of the Wronskian $W(f_1, \ldots, f_n)$.

b) For $\sigma \in \mathrm{Gal}^D(L/K)$, compute $\sigma(W(f_1, \ldots, f_n))$ in terms of the image of σ in $\mathrm{GL}_n(C)$.

c) Generalize to the case of a differential system $Y' = AY$.

Exercise 6.6. **a)** Show that the functions $\exp(x)/x$, $\exp(\exp(x))$ have no elementary antiderivative.

b) Show that $1/(x^2 + 1)$ has no antiderivative in any elementary differential extension of $\mathbf{R}(X)$, but has one in an elementary extension of $\mathbf{C}(X)$.

c) Show that $\sin(x)/x$ has no elementary antiderivative.

Examination problems

I conclude this book by a list of exercises and problems that students were really asked to solve for exams. Some of them are quite substantial.

Exercise 7.1 proves a theorem of Selmer which shows that for any integer $n \geqslant 2$, the polynomial $X^n - X - 1$ is irreducible over $\mathbf{Q}$.

Exercise 7.2 is an elaboration on the casus irreductibilis*: it explains and generalizes the fact that although the three roots of a polynomial equation of degree 3 might be real numbers, Cardan's formulae use complex numbers.*

Exercise 7.5 is a theorem due to Galois about solvability by radicals of equations of prime degree.

Exercise 7.11 proves a theorem due to E. Artin and O. Schreier about subfields F of an algebraically closed field Ω such that $[\Omega : F]$ is finite.

Exercise 7.1. Let n be any integer $\geqslant 2$, and let $S = X^n - X - 1$. You shall show, following E. Selmer[1], that S is irreducible in $\mathbf{Z}[X]$.

a) Show that S has n distinct roots in $\mathbf{C}$.

b) For any polyomial $P \in \mathbf{C}[X]$ such that $P(0) \neq 0$, set

$$\varphi(P) = \sum_{j=1}^{m} \left(z_j - \frac{1}{z_j}\right),$$

where $z_1, \ldots, z_m$ are the complex roots of P, repeated according to their multiplicities.

Compute $\varphi(P)$ in terms of the coefficients of P. Compute $\varphi(S)$.

If P and Q are two polynomials in $\mathbf{Q}[X]$ with $P(0)Q(0) \neq 0$, show that $\varphi(PQ) = \varphi(P) + \varphi(Q)$.

c) For any root z of S, establish the inequality

---

[1] *Math. Scand.* **4** (1956), p. 287–302

$$2\Re\left(z - \frac{1}{z}\right) > \frac{1}{|z|^2} - 1.$$

(Set $z = re^{i\theta}$ and evaluate $\cos(\theta)$ in terms of r.)

d) If $x_1, \ldots, x_m$ are positive real numbers with $\prod_{j=1}^{m} x_j = 1$, show that $\sum_{j=1}^{m} x_j \geqslant m$.

e) Let P and Q be two polynomials in $\mathbf{Z}[X]$, of positive degrees, such that $S = PQ$. Show that $|P(0)| = 1$ and that $\varphi(P)$ is a positive integer. Derive a contradiction, hence that S is an irreducible polynomial in $\mathbf{Z}[X]$.

Exercise 7.2. Let E be any subfield of the field $\mathbf{R}$ of real numbers. By definition, a *real radical extension* of E is a radical extension of E contained in $\mathbf{R}$.

a) Let $E \subset F$ be any finite Galois extension, with $F \subset \mathbf{R}$. Let $\alpha \in \mathbf{R}$ such that $\alpha^N \in E$, where $N \geqslant 2$ is an integer, so that the extension $E \subset E(\alpha)$ is real radical, of exponent N.

(*i*) Let $m = [E(\alpha) : F \cap E(\alpha)]$ and set $\beta = \alpha^m$. Show that β belongs to $F \cap E(\alpha)$. Deduce that $F \cap E(\alpha) = E(\beta)$.

(*ii*) Observe that some power of β belongs to E and show that $[E(\beta) : E]$ equals 1 or 2.

b) Let $E \subset F$ be a finite Galois extension, with $F \subset \mathbf{R}$; let $E \subset K$ be any real radical extension. Show that $[K \cap F : E]$ is a power of 2. (By induction: introduce $L \subset K$ such that $E \subset L$ is elementary radical, apply the induction hypothesis to the Galois extension $L \subset FL$ and to the radical extension $L \subset K$.)

c) Let $P \in \mathbf{Q}[X]$ be an irreducible polynomial of degree n, all the roots of which are real. Assume that one of the roots α of P belongs to some real radical extension of $\mathbf{Q}$.

Show that n is a power of 2.

Exercise 7.3. a) Show that the real numbers 1, $\sqrt{2}$ and $\sqrt{5}$ are linearly independent over the field $\mathbf{Q}$ of rational numbers.

b) We denote by K the field generated in $\mathbf{R}$ by $\sqrt{2}$ and $\sqrt{5}$. What is the degree of the extension $\mathbf{Q} \subset K$?

c) Show that the extension $\mathbf{Q} \subset K$ is Galois and compute its Galois group.

d) Find an element $\alpha \in K$ such that $K = \mathbf{Q}(\alpha)$.

e) Give the list of all subfields of K.

Exercise 7.4. **a)** Let K be a field of characteristic zero, let $d \in K$ such that d is not a square, and let $K \subset K(\sqrt{d})$ be the quadratic extension generated by a square root of d in an algebraic closure of K.

Let $x \in K$. Show that x is a square in $K(\sqrt{d})$ if and only if, either x is a square in K, or dx is a square in K.

Through the end of this exercise, we shall consider three rational numbers r, s and t. Assume that t is not a square in $\mathbf{Q}$. Denote by $\sqrt{t}$ one of the two complex square roots of t and set $E = \mathbf{Q}(\sqrt{t})$. Assume moreover that $r + s\sqrt{t}$ is not a square in $\mathbf{Q}(\sqrt{t})$, denote by α one of the two complex square roots of $r + s\sqrt{t}$, and set $F = E(\alpha)$.

b) Show that $[F : E] = 4$. What is the minimal polynomial over $\mathbf{Q}$ of α? What are its conjugates in $\mathbf{C}$?

c) Show that the following are equivalent:

(i) the extension $\mathbf{Q} \subset F$ is Galois;

(ii) $r - s\sqrt{t}$ is a square in F;

(iii) there exists $x \in \mathbf{Q}$ such that $r^2 - s^2t = x^2$ (first case) or $r^2 - s^2t = tx^2$ (second case).

d) Assume that the extension $\mathbf{Q} \subset F$ is Galois. In the first case, show that $\mathrm{Gal}(F/\mathbf{Q}) \simeq (\mathbf{Z}/2\mathbf{Z})^2$. In the second case, show that $\mathrm{Gal}(F/\mathbf{Q}) \simeq \mathbf{Z}/4\mathbf{Z}$. (Notice that an element of the Galois group is determined by the image of α, at least if $s \neq 0$, and be methodic.)

c) (*Numerical application*) Write $\sqrt{5 + \sqrt{21}}$ without nested radicals. Is it possible for $\sqrt{7 + 2\sqrt{5}}$?

Exercise 7.5. The aim of this exercise is to prove Proposition VIII of Galois's dissertation: "*For an irreducible equation of prime degree to be solvable by radicals, it is* necessary *and* sufficient *that, any two of its roots being known, the others can be deduced rationally from them.*" (French: "Pour qu'une équation irréductible de degré premier soit soluble par radicaux, il *faut* et il *suffit* que deux quelconques des racines étant connues, les autres s'en déduisent rationnellement.")

Part 1. Group theory

a) Let X be a finite set, G a subgroup of the group $\mathfrak{S}(X)$ of permutations of X. Assume that G acts transitively on X. Let H be a normal subgroup in G.

(i) If x and y are two elements in X, and if $\mathrm{Stab}_H(x)$ and $\mathrm{Stab}_H(y)$ denote their stabilizers in H, show that there exists $g \in G$ such that $g\,\mathrm{Stab}_H(y)g^{-1} = \mathrm{Stab}_H(x)$. Deduce that the orbits of x and y under the action of H have the same cardinality.

(*ii*) We moreover assume that the cardinality of X is a prime number. If $H \neq \{1\}$, show that H acts transitively on X.

b) Let p be a a prime number, and let B_p be the subgroup of permutations of $\mathbf{Z}/p\mathbf{Z}$ that have the form $m \mapsto am + b$ with a and b in $\mathbf{Z}/p\mathbf{Z}$.

(*i*) What is the cardinality of B_p?

(*ii*) Show that B_p acts transitively on $\mathbf{Z}/p\mathbf{Z}$. Let h be any element in B_p which fixes two distinct elements of $\mathbf{Z}/p\mathbf{Z}$. Show that $h = \mathrm{id}$.

(*iii*) Show that B_p is solvable. (You might want to introduce the subgroup of B_p consisting of permutations of the form $m \mapsto m + b$, with $b \in \mathbf{Z}/p\mathbf{Z}$.)

c) Let p be a prime number and let G be a subgroup of the symmetric group $\mathfrak{S}_p$. Assume that G is solvable and acts transitively on $\{1, \ldots, p\}$. Let

$$\{1\} = G_0 \subsetneq G_1 \subsetneq \cdots \subsetneq G_m = G$$

be a series of subgroups of G, where for each i, G_i is normal in G_{i+1}, with G_{i+1}/G_i a cyclic group of prime order.

(*i*) Show that G_1 is generated by some circular permutation of order p.

(*ii*) Show that there is $\sigma \in \mathfrak{S}_p$ such that $\sigma^{-1}G_1\sigma$ is generated by the circular permutation $(1, 2, \ldots, p)$.

(*iii*) Identify the set $\{1, \ldots, p\}$ with $\mathbf{Z}/p\mathbf{Z}$ by associating to an integer its class modulo p. This identifies the group $\mathfrak{S}_p$ of permutations of $\{1, \ldots, p\}$ with the group of permutations of $\mathbf{Z}/p\mathbf{Z}$. In particular, the group B_p of question b) is now viewed as a subgroup of $\mathfrak{S}_p$.

Show that $\sigma^{-1}G\sigma$ is contained in B_p.

Part 2. Field extensions

Let E be a field, let $P \in E[X]$ be an irreducible separable polynomial of prime degree p, and let $E \subset F$ be a splitting extension of P.

a) (*i*) Explain how the Galois group $\mathrm{Gal}(F/E)$ can be identified with a transitive subgroup of $\mathfrak{S}_p$.

(*ii*) If $\mathrm{Gal}(F/E)$ is solvable, show that, α and $\beta \in F$ being any two distinct roots of P, then $F = E(\alpha, \beta)$.

b) (*i*) Show that $\mathrm{Gal}(F/E)$ contains a circular permutation of order p.

(*ii*) Assume that there is a root $\alpha \in F$ of P such that $F = E(\alpha)$. Show that $\mathrm{Gal}(F/E) \simeq \mathbf{Z}/p\mathbf{Z}$.

Assume for the rest of this question that there are two distinct roots of P, α and $\beta \in F$, such that $F = E(\alpha, \beta)$.

(*iii*) Show that $[F : E] \leqslant p(p-1)$.

(*iv*) Show that $\mathrm{Gal}(F/E)$ contains a circular permutation of order p.

(*v*) Let σ and τ be two circular permutations of order p in $\mathrm{Gal}(F/E)$. Show that they generate the same cyclic subgroup of order p, and conclude that this subgroup is normal in G.

(*vi*) Show that $\mathrm{Gal}(F/E)$ is solvable.

Exercise 7.6. **a)** Which of the following complex numbers are algebraic numbers? Give their minimal polynomial.

$\sqrt{2}$, $\sqrt{1+\sqrt{2}}$, $(1+\pi^2)/(1-\pi)$.

b) Does there exist $x \in \mathbf{Q}(\sqrt{3})$ whose square is 2?

c) Let $K \subset E$ be a quadratic extension ($[E:K]=2$). Let $P \in K[X]$ be any polynomial of degree 3. One assumes that P has a root in E. Show that P has a root in K. Is the polynomial P split in K? In E?

Exercise 7.7. **a)** Let G denote the finite group $\mathfrak{S}_4$. Let H be the set of permutations $\sigma \in G$ such that $\sigma(1) = 1$.

(*i*) Compute $\operatorname{card} G$, $\operatorname{card} H$. Show that H is a subgroup of G and compute $(G:H)$.

(*ii*) Show that H is not normal in G.

(*iii*) For any $\tau \in G \setminus H$, show that G is generated by H and τ.

b) Show that any group of cardinality equal to 4 contains an element of order 2. Of which general theorem is this a (very) particular case?

c) Let G be a finite group, let H be a normal subgroup of G such that $(G:H)=4$. Show that there exists a subgroup K in G such that $H \subset K$ and $(G:K)=2$.

d) Let $E \subset F$ be an extension of fields with $[F:E]=4$. If it is Galois, show that there exists a field K such that $E \subsetneq K \subsetneq F$. Give an example where such a field does not exist.

Exercise 7.8. **a)** Let E be an infinite field and let $P \in E[X_1, \dots, X_d]$ be a nonzero polynomial in d variables. Show by induction on d that there exists $(x_1, \dots, x_d) \in E^d$ such that $P(x_1, \dots, x_d) \neq 0$.

Let A and B be two matrices in $M_n(E)$. Assume that there is a finite extension F of E such that A and B are conjugate in $M_n(F)$. (Reminder: this means that there is an invertible matrix $P \in \mathrm{GL}_n(F)$ such that $B = P^{-1}AP$.) Denote by $(\alpha_1, \dots, \alpha_d)$ a basis of F as an E-vector space.

b) Show that there are matrices $P_1, \dots, P_d \in M_n(E)$ such that $P_iA = BP_i$ for any $i \in \{1, \dots, d\}$ and such that $\det(\sum \alpha_i P_i) \neq 0$.

c) Show that there are $x_1, \dots, x_d \in E$ such that the matrix $P = \sum_{i=1}^{d} x_i P_i$ is invertible in $M_n(E)$.

Conclude that A and B are conjugate in $M_n(E)$.

Exercise 7.9. **a)** What are the conjugates of $\sqrt{2+\sqrt{5}}$ over $\mathbf{Q}$?

b) Let $K = \mathbf{Q}(\sqrt{2+\sqrt{5}})$. Determine the degree of the extension $\mathbf{Q} \subset K$. Show that this extension is not Galois.

c) Let $\mathbf{Q} \subset L$ be its Galois closure inside the field of complex numbers. Compute $[L : \mathbf{Q}]$.

Exercise 7.10. Let $p_1, \dots, p_n$ be distinct prime numbers. Let $K = \mathbf{Q}(\sqrt{p_1}, \dots, \sqrt{p_n})$ be the field generated in $\mathbf{R}$ by their square roots.

a) Show that the extension $\mathbf{Q} \subset K$ is Galois.

b) Let $\sigma \in \operatorname{Gal}(K/\mathbf{Q})$. Show that there exists, for any $i \in \{1, \dots, n\}$, an element $\varepsilon_i(\sigma) \in \{\pm 1\}$ such that

$$\sigma(\sqrt{p_i}) = \varepsilon_i(\sigma)\sqrt{p_i}.$$

Show that the map ε defined by $\varepsilon(\sigma) = (\varepsilon_1(\sigma), \dots, \varepsilon_n(\sigma))$ is a morphism of groups from $\operatorname{Gal}(K/\mathbf{Q})$ to $\{\pm 1\}^n$.

c) Show that ε is injective.

d) For any nonempty subset $I \subset \{1, \dots, n\}$, show that there is $\sigma \in \operatorname{Gal}(K/\mathbf{Q})$ such that $\prod_{i \in I} \varepsilon_i(\sigma) = -1$. Conclude that ε is surjective.

e) What is $[K : \mathbf{Q}]$ equal to? Show in particular that the real numbers $\sqrt{p_1}, \dots, \sqrt{p_n}$ are linearly independent over the field of rational numbers.

Exercise 7.11. The goal of this exercise is to prove[2] a theorem due to E. Artin and O. Schreier (1927) which characterizes fields F the algebraic closure of which satisfies $[\Omega : F] < \infty$. In particular, one will find that $[\Omega : F] = 2$. Parts 1 and 2 are independent. Part 3 uses results from Part 2.

Part 1. Real closed fields

Say a field K is *real closed* if it satisfies the three following properties:

a) -1 is not a square in K;
b) any element of K is either a square, or minus a square;
c) the field $K(\sqrt{-1})$ is algebraically closed.

a) Show that $\mathbf{R}$ is real closed, as is the the subfield $\mathbf{R} \cap \overline{\mathbf{Q}}$ of real numbers that are algebraic over $\mathbf{Q}$.

b) If K is a real closed field, show that any polynomial of odd degree with coefficients in K has a root in K. Describe more generally the irreducible polynomials with coefficients in a real closed field.

---

[2] This proof is due to W. Waterhouse, *Amer. Math. Monthly,* April 1985, p. 270–273.

c) Let K be a real closed field. If a and b are elements of K, show that a^2+b^2 is a square in K. Prove that any sum of squares in K is still a square.

d) Show that a real closed field has characteristic 0. (Show that for any integer $n \geqslant 0$, $n1_K$ is a square in K.)

Part 2. Cyclic extensions

Let p be any prime number. Let $E \subset F$ be a Galois extension with group $\mathbf{Z}/p\mathbf{Z}$. Denote by σ a generator of $\mathrm{Gal}(F/E)$.

a) If $x \in F$, set

$$N(x) = x\sigma(x)\ldots\sigma^{p-1}(x) \quad \text{and} \quad T(x) = x + \sigma(x) + \cdots + \sigma^{p-1}(x).$$

(i) Show that these are elements of E.

(ii) Show that, for x, $y \in F$, one has $N(xy) = N(x)N(y)$ et $T(x+y) = T(x) + T(y)$.

(iii) Show that, for $a \in E$ and $x \in F$, one has $N(ax) = a^p N(x)$ and $T(ax) = aT(x)$.

b) Assume that there is $\zeta \in E$, $\zeta \neq 1$, such that $\zeta^p = 1$. Show the existence of $y \in F$ with $\sigma(y) = \zeta y$; conclude that $y^p \in E$ and $F = E(y)$.

c) Why does there exist $\theta \in F$ with $F = E[\theta]$? Show that the determinant

$$\det \begin{pmatrix} 1 & \theta & \theta^2 & \ldots & \theta^{p-1} \\ 1 & \sigma(\theta) & \sigma(\theta^2) & \ldots & \sigma(\theta^{p-1}) \\ \vdots & & & \vdots & \\ 1 & \sigma^{n-1}(\theta) & \sigma^{p-1}(\theta^2) & \ldots & \sigma^{p-1}(\theta^{p-1}) \end{pmatrix}$$

is not zero.

d) Show that $T\colon F \to E$ is an E-linear surjective map. Show that for $x \in F$, $T(x) = 0$ if and only if there is $y \in F$ with $x = \sigma(y) - y$.

e) Assume that the field E has characteristic p. Show that there is $y \in F$ with $\sigma(y) = y + 1$, then show that $F = E(y)$ and $y^p - y \in E$.

Part 3. Proof of Artin–Schreier's theorem

Let Ω be an algebraically closed field and let $F \subset \Omega$ be any subfield, with $F \neq \Omega$, such that $[\Omega : F] < \infty$. Denote by i an element of Ω such that $i^2 = -1$.

a) Show that the extension $F \subset \Omega$ is Galois. (If F has characteristic $p > 0$, first prove that F is perfect.)

In the following questions b) to e), assume that $[\Omega : F]$ is a prime number p.

b) What is the Galois group of the extension $F \subset \Omega$?

c) In this question, assume that the characteristic of F is equal to p.

(*i*) If $z \in \Omega$, show that $T(z)^p - T(z) = T(z^p - z)$.

(*ii*) Show that any element of F has the form $x^p - x$, for some $x \in F$.

(*iii*) Using question e) of Part 2, derive a contradiction.

d) Assume that p is odd. Show that there is an element $\zeta \in F$, $\zeta \neq 1$, such that $\zeta^p = 1$. By Part 2, question b), there is $y \in \Omega$ such that $\Omega = F(y)$ and $y^p \in F$. Let $z \in \Omega$ be such $z^p = y$; show that $N(z)^p = y^p$ and obtain a contradiction.

e) It follows that $p = 2$. Show that $\Omega = F(i)$, and that F is a real closed field.

f) (*Back to the general case.*) Show that F is a real closed field and that $[\Omega : F] = 2$.

References

1. E. Artin – *Galois theory*, second ed., Dover Publications Inc., 1998, Edited and with a supplemental chapter by Arthur N. Milgram.
2. M. Audin – *Les systèmes hamiltoniens et leur intégrabilité*, Cours spécialisés, vol. 8, SMF, EDP Sciences, 2001.
3. N. Bourbaki – *Éléments d'histoire des mathématiques*, Masson, 1984.
4. A. Dahan & J. Peiffer – *Une histoire des mathématiques. Routes et dédales*, Points Seuil, 1985.
5. N. Hungerbühler – "A short elementary proof of the Mohr-Mascheroni theorem," *Amer. Math. Monthly* **101** (1994), no. 8, pp. 784–787.
6. I. Kaplansky – *An introduction to differential algebra*, 2nd ed., Publications de l'Institut de Mathématique de l'Université de Nancago, vol. 5, Hermann, Paris, 1976.
7. S. Lang – *Algebra*, third ed., Graduate Texts in Mathematics, vol. 211, Springer-Verlag, New York, 2002.
8. A. R. Magid – *Lectures on differential Galois theory*, University Lecture Series, vol. 7, Amer. Math. Soc., 1994.
9. Platon – *La République*, GF, Flammarion, 1966, introduction, traduction et notes par R. Baccou.
10. M. Rosenlicht – "Integration in finite terms," *American Math. Monthly* **79** (1972), pp. 963–972.
11. W. Rudin – *Real and complex analysis*, 3 ed., McGraw-Hill, 1987.
12. I. Stewart – *Galois theory*, 2nd ed., Chapman and Hall Ltd., 1989.
13. R. Wilson – *Stamping through mathematics*, Springer Verlag, 2001.
14. Zacharie – *Traité du compas. Traité élémentaire de tous les traits servant aux Arts et Métiers et à la construction des Bâtiments*, 1833, available on the Web at Url *http://melusine.eu.org/syracuse/metapost/compas.pdf*.

Index

Undergraduate Texts in Mathematics

(continued from page ii)

Undergraduate Texts in Mathematics

Prenowitz/Jantosciak: Join Geometries.
Priestley: Calculus: A Liberal Art. Second edition.
Protter/Morrey: A First Course in Real Analysis. Second edition.
Protter/Morrey: Intermediate Calculus. Second edition.
Pugh: Real Mathematical Analysis.
Roman: An Introduction to Coding and Information Theory.
Roman: Introduction to the Mathematics of Finance: From Risk Management to Options Pricing.
Ross: Differential Equations: An Introduction with Mathematica®. Second edition.
Ross: Elementary Analysis: The Theory of Calculus.
Samuel: Projective Geometry. *Readings in Mathematics.*
Saxe: Beginning Functional Analysis
Scharlau/Opolka: From Fermat to Minkowski.
Schiff: The Laplace Transform: Theory and Applications.
Sethuraman: Rings, Fields, and Vector Spaces: An Approach to Geometric Constructability.
Sigler: Algebra.
Silverman/Tate: Rational Points on Elliptic Curves.
Simmonds: A Brief on Tensor Analysis. Second edition.
Singer: Geometry: Plane and Fancy.
Singer/Thorpe: Lecture Notes on Elementary Topology and Geometry.
Smith: Linear Algebra. Third edition.
Smith: Primer of Modern Analysis. Second edition.
Stanton/White: Constructive Combinatorics.
Stillwell: Elements of Algebra: Geometry, Numbers, Equations.
Stillwell: Elements of Number Theory.
Stillwell: Mathematics and Its History. Second edition.
Stillwell: Numbers and Geometry. *Readings in Mathematics.*
Strayer: Linear Programming and Its Applications.
Toth: Glimpses of Algebra and Geometry. Second Edition. *Readings in Mathematics.*
Troutman: Variational Calculus and Optimal Control. Second edition.
Valenza: Linear Algebra: An Introduction to Abstract Mathematics.
Whyburn/Duda: Dynamic Topology.
Wilson: Much Ado About Calculus.